NEVER
BY
AUTHORITY

NEVER BY AUTHORITY

WHY REASON DISTRUSTS DOGMA

SERGE BOURGAINE

ORBIS UNUM PRESS

Paleo-Hebrew, the alphabet that appears on the front cover,
was used in the kingdoms of Israel and Judea
during the first millennium BCE.

Library of Congress Control Number: 2020908321

ISBN 978-0-9994906-0-0 (ebook)
ISBN 978-0-9994906-7-9 (print book)

βορyηv01ρβ

To Nicole, as a guidepost on her journey through life.

The universe, and our place in it,
can be explained by observation and experiment,
without resorting to mythical and supernatural explanations.

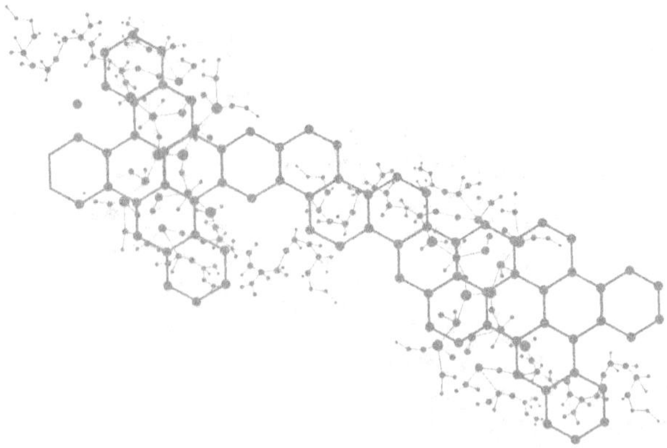

Contents

Preface 1

I. Prologue 9
II. The Origin of Religion 15
III. The Origin of Reason 55
IV. The Cases 87
V. The Case of Gravity 93
VI. The Case of Earth's Age 141
VII. The Case of Human Origins 171
VIII. Belief 209
IX. Perspectives 249
X. Epilogue 283

About The Author 293
Postscript 295
Endnotes 313
Works Cited 323

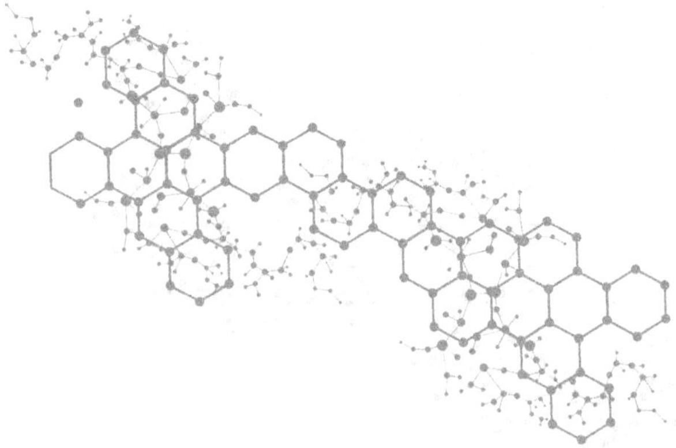

Preface

For what is Truth?
In matters of religion, it is simply the opinion that has survived.[1]
OSCAR WILDE (1854–1900) IRISH PLAYWRIGHT

My daughter inspired this book by way of asking a simple question. "Daddy, what do we believe?" She had just begun to measure her chronological age in double digits. She was curious. She had noticed that many of her childhood friends attended church each Sunday. She had heard words like *God, Lord, Sabbath,* and *Savior*. On several occasions, she had attended church-affiliated summer camps. She was aware of religious references in films, television shows, YouTube videos, pop songs, and books. Her closest friends included Christians, Muslims, and Hindus. I should have anticipated her question. It was a simple question, but I was not prepared with an equally simple answer.

Why was I unprepared to offer a clear response to a legitimate inquiry from an inquisitive child? Perhaps the cursory explanation was that we attended no church and followed no organized religion. Accordingly, I was able to offer no perfunctory *Jesus loves you* or *Praise be to Allāh* declaration, no tale of Yahweh creating the world in six days, no interpretation of the oneness of Ik Onkar, no parable of an invisible entity who watches over you, and no guidance on the spiritual constructs of karma or dharma. I was unable to offer a faith-based answer for the simple reason that we are Enlightenment humanists: those who champion a worldview based on reason; who ground their ethical decisions on classical liberalism; and who reject submission to supernatural authority.

The more substantive explanation, however, relates to the central element of Enlightenment humanism: reason. It is a simple matter to teach ethical values to a young child. Be a nice person, treat others with kindness, harm no one. It is decidedly more difficult to teach reason, which requires a capacity for critical thought. Batson, among other sources, suggests that the ability for critical thinking begins around age eleven.[2] This is the earliest most children are able to grasp the constructs of reason, pose logical questions with at least a nominal degree of lucidity, and understand explanations in a reason-based context. Attempting to teach reason at an earlier age often deteriorates into a rote exercise, one that imparts negligible comprehension. Does this mean no elements of reason can be introduced at an earlier age? No, but it denotes that a comprehensive treatment must await adolescence.

Regardless of explanation, the inability to provide a clear and concise answer to my daughter's inquiry prompted me to revisit my own journey to Enlightenment humanism. I select the word *journey* with intent, for it was a journey. There was no flash of revelation, no instant of illuminating epiphany, and no manifestation from a mythical entity. It was a multi-year odyssey to identify alternatives to faith-based answers which, from adolescence, I invariably found wanting. I elect to cite several formative episodes that epitomize the journey, which I warn the reader was decidedly uneventful.

I was raised in the Presbyterian Church, a branch of Protestantism. As a child, I attended church each week, sat in pews next to fellow worshippers, bowed my head when instructed, drank the grape juice representing the blood of Christ, consumed the bread (now gluten-free in some locales), sang the hymns, and listened to the sermons, all under frequent admonishments from my mother to cease wriggling. I attended Sunday school, learned about Jesus, completed color-by-number pictures from the Bible, and built a paper model of Noah's Ark. The minister said God loved me and would save me, although as a child it was unclear from what I was to be saved.

As I matured, I began to wonder about the teachings of the Church. I noticed a strong emphasis on repetition, a reliance on magic-like phenomena, and the unquestioned acceptance of unexplained events. Could someone, in this instance Lazarus of Bethany, really be dead for several days, undergo the associated natural bodily decay, then be brought back to life?[3] Why does the teaching of this miraculous event omit numerous relevant details? How was this "raising of the dead" accomplished? How was cellular decay reversed? How was desiccated blood reconstituted? Why did it occur only one time? How do we know it really happened? I found such questions germane, as detailed answers could help substantiate the truth of the alleged event. Yet the minister could offer no insights that would ratify such unique and remarkable claims.

As my wonderment grew, I recall one of my first encounters with what I regarded as impaired logic. It concerned predestination. Many branches of Presbyterianism follow the doctrine of John Calvin, an early Protestant reformer. Calvin expounded the doctrine of predestination, which proclaims God as the sole arbiter of salvation. More important, God predisposes the soul of each individual to be saved or to suffer eternal damnation. Yet I recall the minister of our congregation routinely encouraging us to live honest and virtuous lives on the path to salvation. My confusion was simple. If the outcome is foreordained, why is our earthly behavior important? Do the most altruistic and philanthropic deeds carry the individual no closer to spiritual redemption than the alternative, a life of miserly or horrendous acts? I recall raising this question at a Bible study session, but receiving only indistinct, unsatisfactory answers. I was surprised at the vagueness of the responses and, equally, that no one seemed to consider the question relevant.

Another episode occurred during my teenage years when Mormon missionaries, members of the Church of Jesus Christ of Latter-Day Saints, knocked on the door of our home. I listened to their teachings over several visits, but remained both unmoved and unconvinced. I distinctly recall asking what I considered pertinent questions, only to hear the missionaries respond with vague or recited answers.

One instance centered on the Twelve Tribes of Israel. Mormonism teaches that several of the tribes sailed to North America in the eighth century BCE. The Book of Mormon, one of four canons of scripture in the Mormon religion, portrays the native peoples of the Americas as descendants from the tribes.[4] I recall asking about the evidence for this belief. Have researchers discovered ancient shipwrecks on North American shores, dating from the appropriate period? Have linguists identified elements of archaic Aramaic in the structure or vernacular of Native American languages? Have archaeologists unearthed Bronze Age shields, pottery, or coins in the Americas? The response from the missionaries was both unsurprising and unconvincing. No, there is no evidence, but if you pray about it, God will reveal the truth. I did. He didn't.

The final episode I choose to cite occurred when I lived in Utah in the 1970s. At that time, many of the local Boy Scouts of America (BSA) troops were affiliated with the Mormon Church. In Mormonism, boys enter the priesthood at age twelve. The church required senior patrol leaders in Mormon-affiliated scout troops to be members of the priesthood, but there was a problem. The Mormon Church did not accept African-Americans into the priesthood, so they were not eligible to become senior patrol leaders. Church policy stemmed from a Biblical reference in the Book of Genesis. When Cain killed his brother Abel, God "set a mark upon Cain."[5] Early Mormon doctrine taught that Cain's mark was black skin. The doctrine was periodically reinforced by assorted church leaders:

> 1849 Church leader Brigham Young (1801-1877) stated, "The Lord had cursed Cain's seed with blackness and prohibited them the Priesthood."[6]

> 1881 Church President John Taylor (1808-1887) further explained that individuals with black skin were allowed to survive the Biblical flood in order to provide the devil with representation on Earth.[7]

> 1951 Church President David O. McKay (1873-1970) averred that the priesthood restriction was not "the declaration of a policy but of a direct commandment from the Lord."[8]

In the mid-1970s, the National Association for the Advancement of Colored People (NAACP) dismissed the above justifications, alleged discrimination, and sued the BSA. The case, which kindled unfavorable publicity for both the BSA and the Mormon Church, was settled privately. In 1978, however, Church President Spencer Kimball (1895-1985) reported receiving a revelation from God, with guidance to lift the restriction on African-Americans entering the priesthood. Mormons greeted President Kimball's announcement with elation and sincerity. My Mormon friends, with bright eyes and wide smiles, expressed joy in the revelation, affirming their readiness to accept their new brethren into the priesthood.

I, too, was amazed, but for a different reason. I wondered about the convenient coincidence among racial lawsuit, adverse publicity, and timely revelation. Was no one else struck by the fortunate concurrence? What evidence supported the revelation? What evidence demonstrated that the mark placed upon Cain was black skin? What evidence demonstrated that any mark was placed upon Cain? What evidence demonstrated that Cain and Abel in truth existed? Why did the revelation occur at this particular moment? Why were the faithful so willing to accept the pronouncement without question? Again, I found no cogent answers to these questions.

The episodes I recount typify many such encounters. In religious inquiry, I noticed an invariable lack of definitive, evidence-based answers to what I considered pertinent questions. Perhaps the Presbyterian minister and others perceived me as difficult, but I was simply seeking responses I could comprehend. As a teen, I probably wanted to believe and probably wanted to fit in, but it simply made no sense.

As realization mounted that the teachings of the Church were decidedly absent of facts, a chance event occurred. I volunteered to assist at a church-sponsored rummage sale, helping to arrange the various items donated to the sale by members of the congregation.

Among the curios and knickknacks, I stumbled upon the *Great Books of the Western World*, a compilation of works by many of the most profound thinkers in Western civilization. The 54-volume set embraced a wide variety of disciplines, from the arts to politics, from history to theology, and across the formal, natural, and social sciences.

The individual works spanned some two millennia, from the ancient Greeks through the Middle Ages, and from the Renaissance up to the early twentieth century. Many of the authors, still today, are recognizable names: among them the philosophers Plato and Aristotle; the historians Plutarch and Gibbon; the astronomers Copernicus and Galilei; the theologians Aquinas and Augustine; the poets Homer and Virgil; the economists Marx and Smith; the playwrights Sophocles and Shakespeare; the mathematicians Euclid and Descartes; the thinkers Kant and Rousseau; and the essayists Hume and Locke. I purchased the *Great Books* on the day of the sale, then consumed their writings over the following eight years.

While many of the texts defended theological tenets, others openly questioned supernatural ontologies. The latter tomes not only challenged faith-based paradigms but, in their stead, championed reason as the basis for establishing an objective, evidence-based worldview. From this first exposure, the *Great Books* set me on a new path, one that before long led me to adopt Enlightenment humanism as both a rational and defendable philosophy of life.

Such were my journey and its destination. I now extend the journey to author this book, a book about Reason, spelled with a capital *R* to denote its role as the core element of humanism. I write the book for my daughter as a teaching aid through her adolescence, as a framework for discerning myth from reality, and as a foundation for cultivating informed perspectives. Equally important, I write the book as a guidepost for her life, to help her understand the universe and her place in it. Lastly, I write the book to answer the question, "Daddy, what do we believe?"

MISCELLANY

This and subsequent chapters close with assorted miscellany. Here are the first nuggets.

Enlightenment humanism draws its principles from the eponymous revolution that burst forth in the seventeenth century. Apart from reason, the classical liberal values that underpin the philosophy include individual liberty, freedom of expression, the rule of law, equal rights before the law, separation of church from state, an independent press, and the recognition that sovereignty resides with the people. Within this text, I treat Enlightenment humanism, secular humanism and humanism as synonyms.

When Arabic words are expressed in the Latin alphabet, various transliteration and transcription systems employ diacritics (marks written above or below a letter) to assist with pronunciation. For simplicity, this text employs only one diacritic, the macron, to denote long vowel sounds. Thus, Qur'ān appears as the name of the Islāmic holy book, and Hadīth refers to the sayings of Muhammad.

The four canons of Mormon scripture consist of *The Bible*; *The Book of Mormon*, containing the writings of ancient prophets; *The Book of Doctrine and Covenants*, a collection of revelations and writings from past Church leaders; and *The Pearl of Great Price*, a selection of materials composed by Joseph Smith, the founder of Mormonism.

Today I would ask how Mormonism explains archaeological, anthropological, and DNA evidence, which indicate the Americas were first populated at least 15,000 years ago from Asia, not from the Middle East.

To complete the story regarding the mark placed on Cain, in 2013 the Mormon Church published a statement condemning all racism, rejecting its previous teachings that black skin is a mark of divine disapproval.

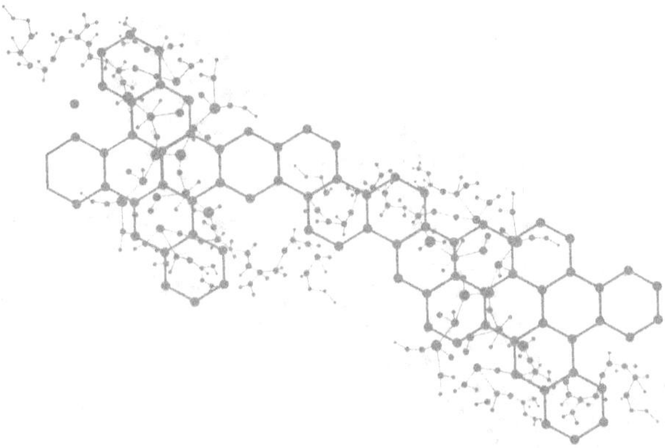

[I]

Prologue

Children must be taught how to think, not what to think.[10]

MARGARET MEAD (1901–1978) AMERICAN ANTHROPOLOGIST

The Royal Society, established in London in 1660, inspired the title of this book. The Society's mission is to promote the advancement of scientific knowledge for the benefit of humanity. Its Latin motto, *Nullius in verba,* commonly is translated as *take nobody's word for it,* or *on the word of no one.* I choose a crisper translation of *never by authority.*

The motto, as articulated on the Society's website, expresses the determination of the Society to "withstand the domination of authority and to verify all statements by an appeal to facts determined by experiment."[11] This statement frames humanity's journey to understand the world we live in based on empirical evidence and testable theories. At the same time, it dispels the use of myth, mysticism, and supernatural explanations in building this understanding.

Nowhere is the motto of the Society more pertinent than in the competition for what we, *Homo sapiens,* think and what we believe. On a personal level, our beliefs drive our actions. On a societal level, our beliefs determine our cultures, our ethics, our institutions, our economic models, our political models, and our ways of life. Multiple disciplines shape these beliefs. Philosophy speculates on our metaphysical origins. History shapes our identities and traditions. Literature reveals insights by way of its prose and poetry. Art and music enlighten through their creativity and creations. Nature inspires from its grandeur and beauty. But two domains reign supreme in the competition for our thoughts, beliefs, and actions: Religion and Reason.

Religion, the more ancient of the two, relies on authority for its dogma. The authority derives from deities and sacred texts, which prescribe the canon and ethos of religious belief. Temporal powers interpret doctrine to guide adherents of the relevant faiths. Reason, in contrast, relies on no authority. Empirical observations and testable predictions are its *sine qua nons*. Reason constructs its principles and laws from experiment, evidence, and independent verification. Unsurprisingly, the dissimilar approaches of Religion and Reason often produce incompatible, even antagonistic, explanations of reality. How can one address the incompatibility? Are the two domains equally valid? Do the pair overlap in some manner? Or do the two realms form a dichotomy, each operating within its own sphere, independent and distinct from its competitor, providing answers only to the inquiries that fall within its boundaries? Is one domain simply right and the other simply wrong? If so, which one is correct?

Operationally, how does each domain arrive at its view of reality? How does each ascertain the validity of a claim? When confronted with one of life's compelling questions, how does each domain ascertain the answer? When an answer is presented, how can one assess whether the response is likely to be true? What is the historical record of each domain in providing correct answers? Which domain should you adopt as your guidepost in life?

To answer these questions, this book advances three principles. First, there is an essential distinction between *how to think* and *what to think*. Second, learning *how to think* is vitally more important than memorizing *what to think*. And third, as to Religion and Reason, one domain is valid, the other is invalid. Let's briefly detail each principle.

PRINCIPLE #1 What is the distinction between *how to think* and *what to think*? *How to think* is the process by which new information is evaluated for acceptance or rejection, or existing information is re-assessed for its validity. Information encompasses assumptions, facts, and beliefs. Applying the explicit steps and techniques of a *how to think* process provides a basis for discerning truth from falsehood. Accepted information is deemed to be true, rejected information false. The former joins the established canon of facts and beliefs that form *what to think.*

PRINCIPLE #2 Why is *how to think* critically more important than *what to think*? Simply put, a sound *how to think* process determines the validity of *what to think.* If the *how to think* process is vague or sloppy, information will be incorrectly assessed. The resulting body of knowledge, the *what to think,* will be inaccurate and misleading. Actions founded on this knowledge are likely to be misguided, even harmful. A rigorous and transparent *how to think* process, by contrast, will more likely produce an accurate compendium of *what to think* knowledge and a reliable set of resulting actions.

PRINCIPLE #3 Which domain is correct, Religion or Reason? Both domains possess *how to think* processes for evaluating salient information and reaching allegedly sound conclusions. Over centuries, each arena has exercised its *how to think* process to generate a vast compendium of *what to think* knowledge, knowledge the domain deems to be true. Yet only Reason is valid; Religion is not. There is no dichotomy, no overlap of domains. Of the two antithetical spheres, only Reason operates an objective and transparent *how to think* process that produces credible, defendable, and independently verifiable *what to think* answers to the large and small questions one encounters in life. This is a bold claim, yet entirely defensible.

This book affirms the above principles in four stages. First, the narrative outlines the origin and development of Religion, then delineates the steps and techniques of its primary *how to think* process, the five sacreds. The text then applies the same treatment to Reason, describing its origin, its evolution, and its principal *how to think* process, the scientific method.

Second, the text examines three specific cases where application of the *how to think* processes of Religion and Reason produced opposing and incompatible *what to think* outcomes: *The Case of Gravity, The Case of Earth's Age,* and *The Case of Human Origins*. All three cases explore the basis and validity of the divergent outcomes.

Third, the narrative investigates the psychological, sociological, and evolutionary origins of religious belief. From this investigation, the book compares the *what to think* catalogs of knowledge from Reason and Religion by providing answers to several of life's most far-reaching questions.

The *Epilogue* closes the text, positioning Reason as the only legitimate guidepost for life. Not just any guidepost, not a guidepost based on myth and superstition, but an objective, practical, and open-minded beacon that promotes the search for knowledge and advocates a humanist worldview.

Let's commence by examining Religion.

MISCELLANY

I elect to capitalize Religion and Reason throughout the text to denote the expanse and import of each domain. Definitions of both terms follow in the next two chapters. For now, it suffices to remark that humans have possessed the ability to reason (lower case *r*) for millennia. Reason (upper case *R*), as defined in chapter three, is a construct based on inductive reasoning and the scientific method. The definition of Religion follows the general understanding of the term.

The three principles expressed in the *Prologue*.

Diagram 1.1

Principles

① There is an important distinction between *how to think* and *what to think*.

② *How to think* is critically more important than *what to think*.

③ Only Reason possesses a sound *how to think* process and a credible *what to think* compendium of knowledge.

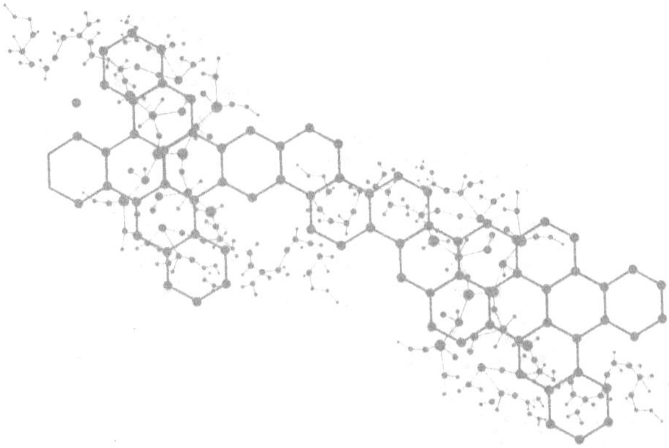

[II]

The Origin Of Religion

In each village in France there is now a lighted torch, the schoolmaster, and a mouth which blows upon it, the priest.[12]

VICTOR HUGO (1802–1885) FRENCH NOVELIST

The year 100,000 BCE

The earth began to shake. Zaug, awakened and startled by the sudden movement, sat upright in his hut. He immediately glanced toward his woman and their two young sons, confirming they were safe. They, too, awoke, instantly alert to potential danger. The boys began to cry as scattered rushes tumbled from the thatched roof, landing by their sides on the earthen floor. The shaking continued for several seconds before stopping as suddenly as it had started. Zaug and his family slowly stood, then cautiously stepped outside into the breaking dawn. Several neighbors in the small, Paleolithic village had emerged from their dwellings, similarly agitated from the movement of the earth.

At the morning meal, Zaug and the village elders discussed the shaking. Khark, the strongest hunter of the band, announced that the migration of a nearby herd of elephants must have caused the movement. Tjang, the village chieftain, disagreed. He declared that a large, nearby tree must have fallen, causing the ground to tremble. Mugh, Bokn, Ogg, and Ubin, the other elders of the band, submitted their views. None, however, offered a conclusive explanation.

While out foraging later that day, members of the tribe kept an eye out for fallen trees and fresh elephant tracks, but saw none. A search revealed no other clues that could account for the earth's movement. They reported this information to the tribal elders at the communal evening meal. After hearing the news, Ogg, a charismatic and ingenious elder, stood and declared, "I know the cause of the shaking. We know there are no fallen trees or elephant tracks nearby. It must, therefore, be a higher, unseen power that caused the shaking. We must try to understand this power and ensure we do not displease it." The other elders considered the words of Ogg, glanced at each other, and nodded. Thus was born Religion.

It is difficult to determine the seedbed of religious belief. Archaeological discoveries worldwide - from ancient dwellings, middens, and burial sites - suggest that spiritual practices emerged tens of millennia ago at sundry locations. Among the unearthed objects are assorted drawings, bones, tools, and stelae. Archaeologists infer spiritual belief from the presence, positioning, and distinguishing features of the objects. The Chauvet Cave in southern France illustrates. Paintings in the cave date from 30,000 years ago, constituting some of the oldest known artwork in the world.[13] Undisturbed patterns on the cave floor and a bear skull positioned on a stone block suggest spiritual habits.

Discoveries from the Qafzeh Cave in Israel boast findings from an even earlier date. Here, burial remains, grave goods, and the presence of ochre-stained bones hint at special rituals dating from 100,000 years ago. Excavations in Spain and elsewhere yield similar findings, affirming the thesis that spiritual practices emerged independently at diverse sites during the Paleolithic Period. If one accepts that modern humans arose approximately 200,000 years ago, then religious belief may have begun soon thereafter. If one interprets religious inclination from the archaeological finds of our now-extinct hominin cousins, the date is older still.

For our narrative, it is useful to establish a start date for Religion. Although its year of origin is unknown, we will adopt half the age of our species as an indicative start date. This yields a date of 100,000 BCE, which aligns not only to archaeological evidence but confirms a Paleolithic genesis of religious belief.

To complement our start date, we likewise will establish a definition of Religion. Most definitions suggest Religion is a structured system of beliefs centered on one or more gods, incorporating both sacred and temporal elements. The former includes deities and sacred texts, while the latter encompasses rituals and rules of conduct. Here are several noteworthy definitions.

A religion is a unified system of beliefs and practices relative to sacred things, that is to say, things set apart and forbidden - beliefs and practices which unite into one single moral community called a Church, all those who adhere to them.[14]

ÉMILE DURKHEIM (1858–1917) FRENCH SOCIOLOGIST

Religion is the human attitude towards a sacred order that includes within it all being - human or otherwise - i.e., a belief in a cosmos, the meaning of which both includes and transcends man.[15]

PETER BERGER (1929–2017) AUSTRIAN-AMERICAN SOCIOLOGIST

Without further ado, then, a religion is (1) a system of symbols which acts to (2) establish powerful, pervasive and long-lasting moods and motivations in men by (3) formulating conceptions of a general order of existence and (4) clothing these conceptions with such an aura of factuality that (5) the moods and motivations seem entirely realistic.[16]

CLIFFORD GEERTZ (1926–2006) AMERICAN ANTHROPOLOGIST

Religion is the sigh of the oppressed, the heart of a heartless world, just as it is the spirit of a spiritless situation. It is the opiate of the people.[17]

KARL MARX (1818–1883) GERMAN ECONOMIST

A broader sampling of definitions yields myriad utilitarian, inspirational, and poetic variations. Yet nearly all embrace a central element that forms the simple definition we will adopt: *belief, in one form or another, in a supernatural power*.

We now have a start date and a definition for Religion. Both will serve as milestones in narrating the journey from proto-beliefs to modern faiths, a journey that was both lengthy and circuitous. Its travelers developed no common itinerary, obeyed no common timetable, and shared no common destination. We can, however, chronicle the odyssey by dividing it into stages, and by selecting a simplifying theme for each stage. We will adopt this approach, describing three notable stages in human history, each marked by a distinct belief system: animism, polytheism, and monotheism. The stages, depicted in Diagram 2.1, serve only as a narrative device for recounting the evolution of religious belief. There is no suggestion that one model of religious belief is superior to another. Indeed, all models abound in today's world, each with millions or billions of adherents.

STAGE 1

The first stage in the journey covers the Paleolithic Period, the span of human history dating from approximately 200,000 to 12,000 years ago. Animism, the oldest type of belief system in the world, characterizes the era. Animism attributes a spirit or spiritual properties to plants, animals, natural phenomena, and other entities, which I collectively will refer to as objects. The spirit forms an intrinsic element of each object, inseparable from its material existence. Animists acknowledge and respect the spirits, often ascribing to them an ability to influence natural events and human behaviors. The spirits, thusly endowed, possess divine powers, which might be expressed benevolently, malevolently, or indifferently.

The list of objects deemed to encompass a spirit is endless: lakes, lightning, and thunder; wind, rocks, and mountains; flowers,

Diagram 2.1
The evolution of religious belief, represented by simplified periods of human history.

Period	Time Period (Years Ago)	Archetypal Belief
Paleolithic	200,000 to 12,000	Animism
Neolithic	12,000 to 2,000	Polytheism
Modern	2,000 to Present	Monotheism

herbs, and trees; lions, tigers, and birds; circles, triangles, and crosses; sunrises, sunsets, and eclipses; stars, moons, and planets; and shadows, stelae, and sacred words. Indeed, some forms of animism assign spiritual qualities to all objects, either as a collection of distinct spirits each inhabiting a specific object, or as a single, universal spirit pervading all objects. How did this spiritual world arise? What drove early humans to adopt animism? Who interpreted the desires of the spirits? Why were the interpretations accepted? To answer these questions, let's return to Zaug.

Consider the world in which Zaug and his tribe lived. For most of human history, individuals were unable to ascertain the origin and operation of the natural forces around them. Why did the ground sometimes shake? Why did a nearby volcano erupt? What was the meaning of the light that arced across the night sky? Why were the rains plentiful one year, yet absent the next? What led one member of the tribe to fall ill, but not another?

It is the nature of our species to want to understand the world around us and, where possible, to control it. Perhaps the volcano erupted because it was instructed to do so. Perhaps the drought was inflicted as a punishment. Perhaps the light in the night sky was an omen. With no other plausible explanations for these phenomena, the more nimble and creative minds developed the idea that higher, unseen powers might be commanding these events. Once embraced by a charismatic and influential leader, the idea quickly gained traction among other members of the tribe or village.

The notion soon led to an epiphany. Maybe we can influence the higher power, thereby exerting a level of control over natural events. Maybe we can better protect ourselves from arbitrary catastrophes. Maybe we can invite the rains to fall more reliably. Maybe we can produce more bountiful harvests. But how do we influence the higher power? By acknowledging its existence, by acting in a way that pleases it, by worshipping it.

It is now that rules of behavior arise that are thought to gratify the higher power. And it is now that the charismatic leader, the tribal shaman perhaps, claims the role of primary intermediary between the village and the higher power. Through a self-accreditation process, the shaman becomes the sole, qualified interpreter of the acts and intent of the spirit. He (or she) might cradle an amulet, read bones, or cast runic stones to discern the desires of the spirit world. Or he might attempt to enter transcendental states and preternatural realms, returning with divine pronouncements, healing remedies, and treatments for the soul. Regardless of technique, the shaman's private conversations with the higher power regulated the behaviors of the entire tribe.

Ogg, now a shaman with prominent status in the tribe, returns from one such self-initiated transcendental encounter and confidently declares, "There is drought because the tribe did not behave in the manner demanded from the higher power, so let us repent. The volcano erupted because our placations were insincere, so let us practice humility. The light in the night sky signaled our oblations were inadequate, so let us construct a larger shrine." Again, the elders considered the words of Ogg, glanced among themselves, and nodded.

The experience of Zaug and his tribe was not unique. As hunter-gatherers, Paleolithic peoples lived in small, often nomadic bands. Each band faced particular survival challenges. Each sought to influence nearby, natural events. A tribe living near a volcano might revere a mountain spirit. A tribe residing in a desert landscape might honor the essence that brings rain. A tribe dwelling in a polar region might cherish the spirit that delivers the warmth of the sun. Their differing circumstances would drive each tribe to develop its own perceptions of life, its own rituals, and its own spiritual practices, all under the cognizance and stewardship of a tribal shaman.

This model discloses two salient points. First, similar animistic themes and conventions emerged among the countless tribes inhabiting the world. And second, poorly-understood phenomena were attributed to unseen spirits. This was common. At this point in prehistory, there simply was no other plausible, explanatory paradigm. There existed no other means to comprehend the world.

Animism was not the only Paleolithic belief system but, arguably, it was the predominant ideology for most of the history of our species, reigning for tens (if not hundreds) of thousands of years. What's more, it laid the foundation for the jump to polytheism, the next stage in our journey.

STAGE 2

Imagine ten hunter-gatherer bands, each with its own sacred practices. The bands settle near or among one another, gradually abandoning their previous nomadic way of life in favor of establishing farms and permanent settlements. The proximity of the bands brings their respective belief systems into competition with each other. Perhaps one of the creeds offers greater intuitive appeal, or perhaps a charismatic leader is better able to champion his or her personal conviction. In this manner, a consolidation of ideologies emerges, with one or two doctrines becoming dominant in the local community, displacing less engaging mythologies. This process repeats itself thousands of times, not only in different places around the globe, but multiple times within a single location as new peoples arrive with dissimilar belief systems.

The preceding description is simplistic but captures the essence of Stage 2 of our journey, the intermingling and merger of early ideologies. The consolidation progressed slowly, bit by bit, over several thousand years, propelled by a seminal event in human history. That event was the rise of agriculture, spawning humanity's first great remaking, a transition to agrarian societies.[18]

The transition began in the early Neolithic Period, a period 8,000 to 12,000 years ago. Scholars commonly point to the Fertile Crescent as its place of origin, an expanse encompassing portions of eight to ten present-day countries, including Egypt, Syria, Lebanon, Israel, and Jordan.[19] It followed closely in the Indus and Huang He valleys, then elsewhere. Let's examine the transition, sometimes referred to as the Neolithic Revolution[20], and its effect on Religion.

While hunter-gatherers relied on hunting and foraging, agrarian societies deliberately planted crops and raised domesticated animals on a broader scale. The transition to a sedentary, agrarian way of life offered several advantages, but two stand out for our story. First, societies achieved greater dependability in year-round food supplies. Crops harvested in autumn could be stored and consumed in winter. Animals fattened in summer could be slaughtered for nourishment as needed. Second, agriculture enabled the production of significantly greater volumes of food, resulting in regular surpluses of foodstuffs.

The greater availability of food supplies catalyzed the transformation of societies. Individuals became more closely tied to the land they farmed, resulting in more stable and permanent communities. Increases in food stores propelled increases in population, both in absolute numbers and in population density. Labour was released from agriculture, as it was no longer necessary for all members of society to engage in food production. The released labor migrated to myriad crafts, creating specialist occupations: bakers, barbers, bowyers, carpenters, chandlers, clerks, cobblers, cooks, coopers, dyers, fletchers, goldsmiths, minstrels, painters, potters, physicians, scribes, soldiers, tailors, tanners, traders, weavers, wainwrights, and untold more.

Food surpluses, larger populations and specialized occupations carry us to the next step in the chain, societal complexity. Land ownership emerged. Deforestation ensued. Cities began to rise. Public architecture sprouted. Political structures expanded. Civil institutions arose. Social interactions surged. Trade burgeoned. Crime mounted. Justice became centralized. Literacy increased. Populations stratified. At the top end, governing classes, literate minorities, military orders, and religious institutions appeared, particularly in urban areas. At the bottom, illiterate peasants and laborers remained tied to the land and to agricultural routines.

Technology likewise advanced. In agriculture via seed selection, soil nutrients, irrigation methods, crop breeding, weed control, and food storage. In writing via the establishment of numerical symbols, alphabets, grammars, lexicons, inks, parchments, codices, and scrolls, providing for the first time a scalable ability to encode and transmit knowledge. In transportation via roads, bridges, canals, sailing ships, and wheeled vehicles, each conveying barrels, crates, and amphorae over previously unthinkable distances. And in war via chariots, catapults, shields, swords, spears, halberds, helmets, and triremes.

Religion evolved in parallel. Ideologies consolidated. Compelling faiths absorbed adherents from tribal creeds. Ecclesiastical orders appeared. Tithing was introduced. Temples were constructed. Festivals sprouted. Moral philosophies emerged. Doctrines were formalized. Religion, across its assorted strains, penetrated all aspects of ancient life: in the Sumerian, Akkadian, and Babylonian empires of Mesopotamia; in the Xia, Qin, and Han dynasties in China; in the Old, Middle, and New Kingdoms of Egypt; in the Athenian, Spartan, and Corinthian city-states of Greece; in the Persian, Etruscan, and Roman societies of the Middle East and southern Europe; in the Olmec, Zapotec, and Norte Chico societies of Latin America; and in the civilizations of the Indus Valley.

Polytheism, the belief in multiple gods, represents the characteristic belief system in Stage 2. Polytheism shares attributes with animism, ascribing higher powers to the same natural phenomena and objects: flora, fauna, mountains, water, weather, planets, and idols. Polytheism, however, assigns a fully-fledged god or goddess to each class of object, each deity wielding nearly boundless authority over its respective remit. Further, polytheism organizes and codifies its deities into pantheons, hierarchies of inter-related gods who collectively administer both the temporal and divine spheres.

Exhibiting the same desire as their animistic predecessors to influence natural events, polytheistic peoples revered and worshiped the pantheons. In the Paleolithic, the tribal shaman Ogg might have attempted to channel the spirit world. Now, his polytheistic descendent invoked the pantheon. To prevent the volcano from erupting, let us pay homage to the mountain god. To banish drought, let us placate the rain god. To ward off evil from the light in the night sky, let us pray to the moon god. To slay the invader, let us worship the god of war.

To differentiate among the multiplicity of gods inherent in polytheism, societies assigned to each deity a name. Ancient Egypt was particularly proficient at naming. Necessity demanded it. To cite a few of their approximately 200 to 2,000 (estimates vary) tutelary deities: Ra, the sun god; Khon, the moon god; Ptah, the god of creation; Isis, the goddess of motherhood; Khem, the god of fertility; Maut, the goddess of nature; Tefnet, the goddess of rain; Bast, the goddess of cats; Thoth, the god of wisdom; Nut, the goddess of the firmament; Anubis, the god of mummification; Osiris, the god of the underworld; Set, the god of chaos; Bes, the god of pregnant women; Seshat, the goddess of writing; and Ammon, the sovereign over all.

So, too, was each god linked to a primary place of worship. Again, a few examples serve: Ra, the city of Heliopolis; Ptah, the city of Memphis; Khem, agrarian areas; and Osiris, the temples at Abydos and Philae. Further, each god was linked to a sacred animal or concept, so yet another representative list: Osiris, a bull; Isis, a cow; Ra, hawks; Horus, lions; Maat, truth; Anubis, wolves; Set, hippopotami; Shu, the wind, and Kneph, rams.[21]

The Romans likewise were polytheists, acknowledging over one hundred gods (again, estimates vary) over the 1,000-year life of their empire from 509 BCE to 476 CE. They created their own private and public gods, while also adopting deities from their conquests in Greece and elsewhere. Among their pantheon, they worshipped Jupiter, king of the gods; Neptune, god of the seas; Juno, goddess of women; Mars, god of war; Bellona, goddess of war; Venus, goddess of love; Minerva, goddess of wisdom; Janus, god of beginnings; and Vesta, goddess of home and hearth. Some of their minor gods reflect the laudable creativity of the Romans: Annona, god of the grain supply for Rome; Somnus, god of sleep; and Bacchus, god of wine. To this day, immoderate toasts to Bacchus continue to summon Somnus.

Religions in Sumerian, Persian, Greek, Maya, Inca, and other ancient civilizations traced a similar pattern to the Egyptians and Romans - polytheistic beliefs attended by a plethora of deities, each governing its own remit. Among all the pagan deities, however, the most common was a sun god. Each religion seems to have recognized one: Ra for the Egyptians; Tonatiuh for the Aztecs; Lugh for the Celts; Apollo for the Greeks; Inta for the Incas; Kinich Ahau for the Maya; Sol Invictus for the Romans; Amaterasu for the Japanese; and Sol for the Norse peoples. Across religions, sun gods tended to be male and moon gods female. Other deities were a mixture of male and female, including war gods.

You now should have the impression that myriad deities emerged in the Neolithic Period. Intermediaries accompanied the deities: shamans, witch-doctors, medicine-men, healers, prophets, sooth-sayers, mediums, diviners, and priests. Consecrated venues buttressed the intermediaries: temples, churches, chapels, shrines, pagodas, oracles, and places of pilgrimage. Royalty, nobility, and patricians sustained the venues through their patronage and funding. Religious hierarchies emerged in concert, with their members mediating the commands of their deities to shepherd their committed and wayward flocks. Thus was born *organized* Religion in the ancient world.

The societal transformations outlined in this section continued throughout the Neolithic and beyond, extending well into the Modern Period. Yet, by around 1000 BCE, a notable psychological change had emerged. Individuals had begun to learn how to govern their environment. They were no longer at the full mercy of nature. A short drought or small flood could be managed. A protracted winter or a cold summer could be survived. Perhaps this raised the first small, minuscule doubt regarding the validity of animism and polytheism. Perhaps some questioned the plethora of deities, each administering a specific phenomenon of nature. Regardless of cause, a new ideology began to emerge, one that challenged the multiplicity of deities and carries us to Stage 3 of our journey.

STAGE 3

The rise of monotheism, the belief in a single god, marks the final stage of our odyssey. The distinction between polytheism and monotheism is not merely an arithmetic imbalance of multiple deities versus one. A qualitative distinction also obtains. Monotheism commonly posits a sole, omniscient, omnipotent, and omnipresent deity as creator of the world and all life upon it. It acknowledges a divine being who prescribes an ethical order, one who regularly intervenes in the natural world to herd its flock to eternal salvation. Thus, the monotheistic deity is a more sweeping, all-encompassing, and all-embracing entity than most polytheistic gods. And rather than co-existing with other gods, the monotheistic deity negated and supplanted its polytheistic competitors. Yet monotheism's place of origin is unclear. Some scholars credit pharaonic Egypt, others the Zoroastrianism of ancient Persia. Still others cite Judaism, a monotheistic religion whose origin lies in the second millennium BCE. Regardless of source, monotheism advanced in the first century BCE with the rise of Christianity, followed by Islām around 600 CE - both flourishing into the two largest religions of today's world.

Monotheism's rise begs a fundamental question. What caused belief in a single deity to arise? One view is that monotheism arose from henotheism, a creed that denotes belief in one god, but does not deny that other, equally valid gods might exist. Perhaps it was a short mental leap from polytheism to henotheism, then another slight jump to monotheism. But again, why take the step at all?

Another view argues the time was ripe for a new belief system, and monotheism just hitched along for the ride. This view suggests that a novel dogma arose, one offering a more appealing set of beliefs. Rather than navigating the confusing and conflicting dictates of multiple gods, monotheism asserted a single god. Rather than enduring the whims of classical gods, who often were cruel and toyed with their human subjects, the new teaching offered a caring, personal god. Rather than advancing a vague conception of life after death, the new doctrine promised an afterworld of eternal paradise. And rather than tolerating perceived iniquities, the new creed prescribed a rigid virtue (in spirit if not always in practice). From these and similar inducements, converts from animism and polytheism favored the new belief, driving its acceptance and ultimate dominance. In this view, monotheism was simply one component of a new gospel, only later occupying a position as a core tenet.

Today, many religions embrace monotheism in full or in part. In addition to Zoroastrianism, Judaism, Christianity, and Islām, other examples include Bahá'í Faith (Middle Eastern origin), Cao Dai (Vietnam), Mormonism (USA), Rastafarianism (Jamaica), and Sikhism (India). One casualty of monotheism, unsurprisingly, was the pantheon of polytheistic deities. Deities who once formed an essential element of society were expunged from the new doctrine and slowly vanished. Mrs. God provides an interesting example. She was the wife of Yahweh, a Hebrew word for god. References to God's wife in the ancient kingdoms of Israel and Judah make an appearance in religious texts dating from 1200 BCE. By 500 BCE, however, she no longer appears as a co-creator. The cause of her disappearance is unclear. Perhaps she fell victim to a male-dominated culture interpreting a new monotheistic religion. Or perhaps, as with innumerable other gods, she simply was a casualty of a pantheistic purge.[22]

It is essential to note that narrative simplicity causes us to omit many other types of belief systems, such as deism, dualism, and indigenous folk religions. All played distinctive roles among their followers at various points in history. Many continue to do so. It also bears repeating that monotheism is not presented as a superior belief system, but merely as a step in the evolution from animistic and polytheistic beliefs. Numerous non-monotheistic belief systems endure today, particularly as Asian, Indian, and African traditional religions. Here are some representative examples. Each boasts more than one million adherents.

> **Confucianism** is a traditional belief system originating in China. Though sometimes referred to as a religion, it more frequently is described as a humanistic philosophy for leading an ethical life that does not derive from a deity.
>
> **Jainism** is a polytheistic religion in India, but rejects a creator deity on the basis that the universe was never created. The universe exists independently and self-sufficiently. Souls, housed in humans and plants, are eternal, and they attain perfection by cleansing bad karma. Jainism emphasizes asceticism and nonviolence.

Shintō is a polytheistic religion in Japan that recognizes gods or spiritual essences (*kami*) in natural objects and phenomena, such as mountains, trees, waterfalls, and fields.

Taoism is both a Chinese philosophy and pantheistic religion that emphasizes the Tao, the *path* or *way*. As the underlying, unifying principle of the universe, the Tao is present in all things. Taoists emphasize simplicity, humility, and living in harmony with the natural flow of life.

Asatru and **Wicca** are polytheistic, pagan religions, established in the last 100-150 years. Although each possesses fewer than one million adherents, I include them as representative examples of a general class of ancient Norse, Celtic, and other pagan religions. Some profess an afterlife (Valhalla), others do not.

Here it is important to remark that nearly all animistic, polytheistic, and monotheistic religions occur in multiple forms, with beliefs that vary by denomination, school, or sect. Royal decrees, purported revelations, and disagreements between religious leaders - among other factors - periodically induce offshoots to arise from a core religion. Occasionally, the offshoots are reabsorbed, but often they secure a sufficient following and thrive as a separate denomination. In the West, the rise of Protestantism under Martin Luther (1483-1546) offers one of the more famous examples.

Martin Luther was a German theologian and priest, born into the Catholic Church. At a young age, he was struck by lightning in a field, prompting him to swear an oath to God. He vowed that if allowed to recover from his injuries, he would become a monk and dedicate his life to God.[23] He recovered, fulfilled his commitment to become a monk, but gradually grew disillusioned with the teachings of the Church.

In 1571 Luther expressed his displeasure by publishing his *Ninety-Five Theses*, challenging the Church's use of indulgences. Indulgences, a type of spiritual get-out-of-jail-free card, could be purchased from the Church, and occasionally from private brokers, to excuse past, present, or future sins. Common sins included legitimizing children born out of wedlock; permitting nuns to maintain servants; marrying outside the laws of sanguinity; and trading with Muslims. Ironically, priests purchased the majority of dispensations to legitimize their own children.[24] Fees from indulgences formed a considerable source of income for the Church, with the counting of cash a common sight in the Apostolic Palace.

In 1520 the leader of the Church, Pope Leo X, demanded that Luther denounce his writings. One year later, in 1521, the Holy Roman Emperor, Charles V, repeated the demand. Luther refused and was promptly excommunicated. This act, however, had little effect. Luther continued to impugn the power of the pope, arguing that salvation derives from faith alone, without the requirement of good deeds. Among other feats, he translated the Bible from Latin into the vernacular, allowing the laity to access Holy Writ directly, without priest, bishop, or cardinal as intermediary. In the 1520s, he set about establishing a new church, launching the Protestant Reformation. Catholic and partisan leaders attempted to check the Reformation through a succession of religious wars in the 1500s and 1600s, most notably the Peasants' War in 1524-1525 and the Thirty Years War from 1618-1648. But it was too late. The Reformation was launched, and Protestantism was firmly established. It would grow into the second largest branch of Christianity in today's world.

The journey from animism to monotheism now brings us to present day, where five religions have grown beyond their geographic origins to emerge as the world's largest. Their method of expansion varied, but commonly involved word-of-mouth, proselytizing, royal decree, migration, conflict, and conquest. The rise and fall of empires, the formation and abrogation of allegiances, and the extension of trade routes assisted their spread. Today's top five appear in Diagram 2.2. Here is a brief description.

Diagram 2.2

The principal deities and number of adherents of today's top five religions. Worldwide numbers vary by source.

Religion	Principal Deity	Adherents
† Christianity	God	2.4 Billion
☪ Islām	Allāh	1.7 Billion
ॐ Hinduism	Brahma, Shiva, Vishnu	1.1 Billion
☸ Buddhism	Buddha	500 Million
☬ Sikhism	Ik Onkar	25 Million

Christianity is a monotheistic religion. Its adherents worship God, who manifests himself as Father, Son, and Holy Ghost. Its sacred book is the Bible. Its earliest tenets are the Ten Commandments, a list of rules Christianity's followers are taught to obey. The religion derives its name from the Son and Savior, Jesus Christ. The Christian god is a personal god.

Islām also is a monotheistic religion. Its adherents worship Allāh and revere the Prophet Muhammad (c.570-632). Its most sacred texts are the Qur'ān and the Hadīth. The basic tenets of Islām encompass the Five Pillars, consisting of *shahādah,* a declaration of faith in one god (Allāh); *salāt,* the performance of five daily prayers; *zakāt,* practicing charity for the needy; *sawm,* fasting throughout the month of Ramadān, the ninth month of the Islāmic calendar; and *hajj,* making at least one pilgrimage to Mecca, the birthplace of Muhammad, during the twelfth month of the Islāmic calendar, Dhū al-Hijjah, and completing nine essential rites.[25] The Islāmic god is a personal god.

Hinduism is perhaps the oldest religion among the top five. It is a polytheistic religion embracing a broad range of philosophies (some sources describe Hinduism as monotheistic, with one supreme god that manifests itself as multiple deities). Its principal deities are Brahma, Shiva, and Vishnu. Its earliest scriptures are the Vedas, Upanishads, Puranas, and Bhagavad-Gītā. Prominent Hindu beliefs focus on ethics and duties (Dharma); actions and consequences (Karma); liberation to a state of bliss (Moksha); paths to attain liberation (Yogas); and the cycle of birth, life, death, and rebirth (Samsāra).

Buddhism likewise encompasses a broad range of philosophies, resting largely on the teachings of Buddha (the Awakened One), Siddhartha Gautama. More than one source describes Buddhism as a humanistic philosophy for leading an ethical way of life, not a religion. For sects following Buddhism as a religion, however, Buddhism is polytheistic. Buddhism does not presume a personal god. Its basic tenets are based on the Four Noble Truths: suffering exists; there is a reason for suffering; there is a way to end suffering; and the way to end suffering and reach nirvana is through the Eightfold Path.

Sikhism is a monotheistic religion. Its deity is Ik Onkar, the One Supreme Reality. Its most hallowed text is the Gurū Granth Sāhib. The teachings and traditions of Sikhism were enshrined through the agency of ten gurus over the fifteenth to eighteenth centuries. The first guru, Guru Nanak Dev, taught three principles: work hard and honestly (Kirat Karo), share with the needy (Wand Ke Chhako), and remember God throughout the day (Naam Jappo).

Here I repeat an earlier caveat. Each religion has myriad variations. The preceding descriptions are general. They fail to capture the unique beliefs and traditions that ofttimes differ by denomination or sect. Christianity, for example, contains multiple branches: Catholics, Lutherans, Anglicans, Calvinists, Baptists, Anabaptists, Pentecostalists, Quakers, Charismatics, Presbyterians, Methodists, Evangelicals, Mormons, Nontrinitarianists, and Orthodox, to name but several. Many of the denominations found expression from the Protestant Reformation. Of the 2.4 billion Christians worldwide, roughly 55% are Catholic, 45% Protestant.

Islām, too, had its "Protestant Reformation" event.[26] When the Prophet Muhammad died in the year 632, the succession plan was unclear. Rival claimants, subsequent wars, and assassinations led to a schism among Sunni (traditionalists), Shia (followers), and Khawarij (secessionists) Muslims. Divergent theological interpretations within each major school of Islām spawned distinct sub-schools. Sunnism, for example, encompasses Malikism, Shafism, Hanafism, Hanbalism, and Wahhabism, while Shiism includes the Twelvers, Seveners, Druze, Alawite, and Zaydis sub-schools.[27] Today, of the 1.7 billion Muslims in the world, approximately 85% are Sunni and 15% are Shia.

Hinduism, Buddhism, and Sikhism likewise incorporate variants. Buddhism can be divided into the Theravada, Mahayana, and Vajrayana schools, each with many sub-schools, while Hinduism contains the Nyaya, Samkhya, Mimamsa, Carvaka, and Yoga schools, here again with a wealth of sub-schools. By embracing a multitude of philosophies, Buddhism and Hinduism probably win the prize on number of variants.

As previously noted, variations within each major religion often arise from differing interpretations of sacred texts. The interpretations lead to differences not only in beliefs, but also in religious practices, traditions, customs, and jurisprudence. To an outsider, the disparities may appear trivial. But to an adherent, they are crucial. In Christianity, for instance, one belief separating Baptists from other Protestant denominations regards baptism. Baptists believe baptism should be performed on professing believers, not on infants, and that baptism should be conducted by means of immersion, not by sprinkling. As an aside, Baptists experienced their own schism in 1845. In that year, Southern Baptists in the United States split from northern Baptists. Proximate causes were the question of whether slave owners could serve as missionaries, and a quarrel over whether slave ownership was sanctioned in the Bible.

It should be added that the beliefs of a particular faith often originated from a prior creed. Indeed, most religions practice syncretism, the adoption of at least some of their beliefs from other theologies. Contemporary Western faiths, as case in point, were influenced by Mesopotamian, Egyptian, Jewish, Greek, and Roman religions, as well as by European pagan faiths. Here is a sample of some of the beliefs and practices that Christianity adopted from its precursors.

- The concept of an underworld.
- The veneration of saints.
- The existence of angels.
- The declaration of a day of judgment.
- The insistence on one or more virgin births.
- The promise of an exclusive path to truth.

- The resurrection of a departed deity.
- The assurance of salvation.
- The association of shrines to holiness.
- And the rebranding of pagan festivals as Christian commemorations.

A specific example of syncretism is the flood myth. Documented accounts from Sumer and Babylon, relating the tale of individuals boarding a vessel to escape a flood, pre-date Christianity. One example appears in the *Epic of Gilgamesh,* a poetic narrative dated to 2100 BCE, commonly cited as the oldest known work of literature in the world. The *Epic,* discovered in 1853, is written in cuneiform on clay tablets. In the tale, the gods warn a man named Utnapishtim of an impending flood, and they instruct him to construct a ship. Utnapishtim takes aboard his family, some craftsmen, and animals. They survive the deluge, while all other life forms perish.[28]

A second case appears in the *Bibliotheca* (*The Library*), a work by Pseudo-Appollodorus that also mentions the flood myth.[29] In this adaptation, the god Zeus becomes enraged, and he decides to eradicate Bronze Age men. His tool of destruction is a flood. Deucalion, the son of Prometheus (a former Titan who created humanity from clay), constructs a chest and floats on it with his wife Pyrrha "for nine days and as many nights." Upon reaching land, he at once sacrifices to Zeus for his good fortune, prompting Zeus to grant him a wish. Deucalion petitions the re-creation of men. Zeus honors the entreaty, assembling nearby stones to be cast by Deucalion and Pyrrha. The rocks thrown by Deucalion become men, those pitched by Pyrrah become women, thus re-populating the world.

Scholars observe that syncretism often aids the adoption of new faiths. Retaining comforting beliefs from earlier theologies eases the mental leap from old faith to new. Acolytes find conversion less abrupt, less perilous. Here, the Christian Trinity is offered as a case in point. In polytheistic societies, a pantheon of deities oversaw all aspects of an individual's life. A potential convert, when asked to transfer his faith to a monotheistic deity, might express reluctance. Swapping a cornucopia of gods for a single entity might seem a poor wager. According to this thesis, the concept of the Trinity arose as an appealing halfway point.[30] By offering one god in three forms, Christianity eased the path to conversion for those individuals who took comfort in a plethora of deities.

Syncretism also assists the decline of older, once popular religions. Here, theologians play a key role in hastening the decline, by distancing newer religions from their predecessors, and by portraying the latter as founts of superstition and false beliefs.

This completes our sketch on the evolution of Religion, from its Paleolithic roots to its present day forms. To recap, we highlighted three stops on the journey: the animistic origin of Religion, its consolidation under polytheism, and its evolution into monotheism. We also briefly described the five largest religions in today's world, noting that each consists of multiple variants. The quintet, accompanied by countless smaller but no less consequential religions, profoundly affects the everyday lives of their billions of followers, a topic we touch on later in this chapter. We now shift our focus to the primary *how to think* process of Religion.

THE FIVE SACREDS

A goal of this book is to compare how Religion and Reason teach their respective adherents *how to think*. This section examines the five sacreds, the principal *how to think* process of Religion. The five sacreds, as the label implies, comprise five elements inherent in most belief systems: sacred, deities, sacred texts, sacred places, sacred rituals, and sacred persons.

DEITIES

The first sacred, a religion's deity or deities, is central to the existence of its faith. Our definition of Religion, the belief in a supernatural power, acknowledges this reality. Beyond the deity's mere existence, however, each religion also ascribes notable traits and ideals

Diagram 2.3
The five sacreds, the principal *how to think* process of Religion.

1. Deities
2. Texts
3. Places
4. Rituals
5. Persons

to its god. One religion may emphasize forgiveness and redemption (Christianity), another may accentuate obedience to divine law (Judaism), still another may stress purification (Shinto), and a fourth may trumpet compassion and kindness (Sikhism). While the traits vary by religion, they typically leverage a mix of the tri-omni attributes commonly imputed to deities: omnipotence, omniscience, and omnipresence. The deities, thusly invested, are revered and worshipped. Their supremacy and dominion are absolute.

How does a deity communicate its wishes? A primary method is revelation, the disclosure of truth or knowledge to selected individuals. The disclosures, at times explicit but ofttimes coded, reflect the wisdom and tutelage of the deity. Once parsed and disseminated, the revelations apportion unquestioned guidance to the faithful.

SACRED TEXTS

Sacred texts contain scripture, writings which document the revelations of a deity. As a god's revealed word, scripture is deemed to impart truth. Every literate religion possesses at least one such text. The testaments typically chronicle the history of the faith and manifest its doctrines. They also embody moral and ethical principles that confer guidance on how to lead one's life in accordance with the religion. Sacred texts include the Bible (Christianity); the Qur'ān and the Hadīth (Islām); the Vedas and Upanishads (Hinduism); the Gurū Granth Sāhib (Sikhism); the Sutras (one of 1,000+ texts in Chinese, Tibetan, Thai, and other Buddhist schools); the writings of Báb and Bahá'u'lláh (Bahá'í Faith); the Five Classics (Confucianism); the Agamas (Jainism); the Torah and Talmud (Judaism); the Popol Vuh (Maya); the Tao Te Ching (Taoism); and the Yasna (Zoroastrianism).

Among the five sacreds, scripture is subservient only to the deity itself. The texts, albeit written or edited by humans, are presumed to be divinely inspired. All sacrosanct books, however, require interpretation, regardless of authorship. Religious leaders and scholars typically conduct the exegesis. The texts also demand translation. Historians, philologists, and linguists assist in this task. The Bible, originally written in Aramaic, Hebrew, and Greek, has been translated into more than 500 languages. Translations of the Qur'ān, originally written in Arabic, exist in at least as many languages. New translations arise periodically to expound modern interpretations or merely to reflect changes in language. In the latter instance, the archaic vocabulary, grammar, syntax, and phraseology of ancient translations sometimes are unclear to present-day readers.

Sacred texts also often include lists of instructions to guide the lives of followers. Witness the 613 commandments in the Torah (Judaism), the Ten Commandments (Christianity), the Five Pillars (Islām), the Four Noble Truths and Eightfold Path (Buddhism), the Five Thieves (Sikhism), the Seven Valleys (Bahá'í Faith), the Threefold Law (Wicca), and the Five Vows (Jainism). Lists serve as a powerful communication method, both within and outside of Religion, conveying a do-this-to-achieve-that assurance.

SACRED PLACES

The third sacred is sacred places, spaces where holy and earthly elements combine: churches and chapels, mosques and basilicas, synagogues and temples, abbeys and shrines. Sacred places mark the boundaries of distinguished or exceptional spaces. The spaces, in turn, serve as communal venues for worship and education, each designed to transmit and reinforce the teachings of the faith. The edifices frequently incorporate symbols of the religion in a unique architectural style. The most sacred site in a religion often is where the religion was established, its founder was born, or its founder ascended to an afterlife. Notable sacred sites include the Dome of the Rock in Jerusalem (Islām); the Church of the Nativity in Bethlehem (Christianity); the Mayadevi Temple in Lumbinī (Buddhism); the Golden Temple in Amritsar (Sikhism); the Wailing Wall in Jerusalem (Judaism); the Jingū shrine in Ise (Shinto); the Temple of Apollo at Delphi (ancient Greeks); the Pyramid of the Sun in Teotihaucán (Aztecs, arguably); and Stonehenge (Druids, also arguably).

Sacred places frequently adorn their spiritual surroundings with hallowed objects. A church in Italy might purport to possess a vial of blood from the Savior or a splinter of wood from the cross of the crucifixion. A synagogue in Israel might commission a Torah scroll. A mosque in Iran might cherish an illuminated book of early Islāmic calligraphy. Similar artifacts, both ancient and contemporary, abound: altars and stelae; garments and headdresses; pendants and amulets; effigies and scrolls; idols and statues; glyphs and relics; and sundry religious imagery. Each reflects a revered facet of a faith. Each attracts the piety of ardent disciples.

Diagram 2.4 lists the most important sacred places and texts for the five largest religions.

SACRED RITUALS

Sacred rituals comprise the fourth sacred. All religions recognize some combination of spiritual, communal, and personal rituals to be performed on a routine basis. In context of *how to think*, the rituals serve as a vehicle to celebrate and bolster the faith's doctrines. Rituals flourish in all shapes and sizes: pilgrimages, consecrations, sacrifices, festivals, feasting, fasting, ancestor worship, and rites of passage. Rituals often are linked to the calendar, to be performed on or near specific dates. The most common ritual is attendance at a daily or weekly religious service.

Intraday rituals also exist, where a devotional act is performed at specific times of the day. One example is the five daily prayer times (*salāt*) laid down by Islām: after dawn and before sunrise (*al-fahr*), between the start of the sun's ascent until late afternoon (*az-zuhr*), between the point when the sun is halfway through its descent until just before sunset (*al-'asr*), between sunset and when the last red glow in the sky has faded (*al-maghrib*), and between dusk and midnight (*al-ishā*).

Catholicism likewise possesses an intraday ritual, the *Liturgy of the Hours*, initially consisting of eight prayer times: matins (predawn), lauds (dawn), prime (early morning), terce (mid-morning), sext (midday), none (mid-afternoon), vespers (evening); and compline (night). The number and times of prayer derive from interpretations of scripture, such as "In the morning I will direct my prayer unto thee"[31] and "Pray without ceasing."[32] Holders of priestly offices, selected religious orders, and some laity observe the liturgi-

Diagram 2.4
The principal sacred texts and sacred places of the world's top five religions.

Religion	Sacred Text	Sacred Place
† Christianity	Bible	Jerusalem
☪ Islām	Qur'ān	Mecca
ॐ Hinduism	Vedas	Varanasi
☸ Buddhism	Sutras	Lumbini
☬ Sikhism	Gurū Granth Sāhib	Amritsar

cal hours. Since their inception, the number of prayer times has evolved. Present-day practitioners observe five to seven iterations.

Less frequent rituals are driven by the movement of a planet, the arrival of a new year, the appearance of a solstice, the birth of a child, or the death of a follower. Sufism, a branch of Islām that arose in the eighth century, provides an interesting example.[33] Some Sufi denominations embrace the use of music, drums and dance in their rituals, working themselves into hypnotic trances under the direction of a Dervish, a religious guide. This is the origin of the phrase Whirling Dervish that came into use during the nineteenth century in the West, but is now archaic.

Whether holy or festive, rituals fortify beliefs and provide an opportunity to commune with fellow followers. Diagram 2.5 highlights some of the prominent communal rituals of the five largest religions.[34] The usual caveats apply. The names, dates, and nature of the rituals vary by denomination and country. Dates also vary based on the calendar that each religion uses, as laid out in the *Postscript*. The diagram reflects dates for the year 2020.

SACRED PERSONS

The fifth and last sacred is sacred persons, individuals who possess a special significance within a religion. They might be founders of a religion, interpreters of its faith, or authors of its sacred texts. They might be self-appointed or elected. They might be living, or they might have passed into the embrace of their deity. During their tenure as a deity's spokesperson on Earth, they tend to share common traits. They passionately articulate the verity of their beliefs. They are more loyal to their truth than to the audience to whom they preach. They are persuasive, often charismatic individuals. They frequently admit their faults. They convey trust, wisdom, and divine inspiration. They aver the truth of scripture and instruct correct moral behavior. And, above all, they speak with the authority of their deity.[35] Their manifold rubrics, irades, benedictions, and blessings impart the word of their god.

Perhaps the most famous example from the ancient world was Pythia, the Oracle at Delphi.[36] Pythia offered prophecies on behalf

Diagram 2.5
Characteristic rituals of the world's top five religions.

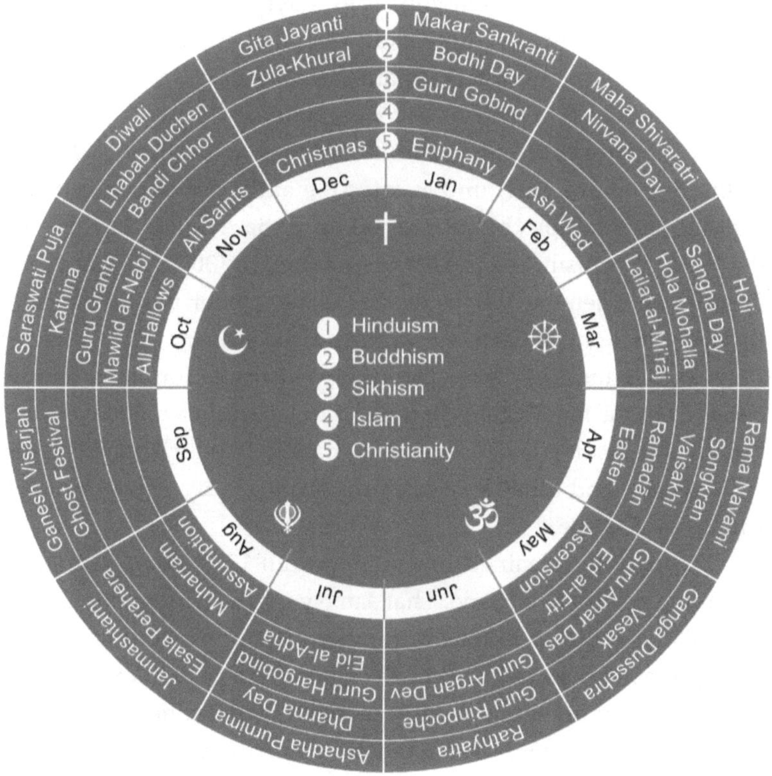

of the Greek god Apollo from the eighth century BCE to the fourth century CE. A modern example is the pope, the head of the Catholic Church. The pope is deemed to be the successor to Saint Peter, who was instructed by Jesus Christ to lead the Church. Another is the Prophet Muhammad, whose revelations from Allāh form the verses of the Qur'ān. A fourth example is Guru Nanak (1469-1539), the founder of Sikhism. Through their writings, teachings, interpretations, and *ex cathedra* pronouncements, sacred personages set the temporal direction of a religion. They often are perceived to communicate directly with a deity and to channel the word of the deity. Workaday priests, ministers, imāms, and other clergy then propagate the messages to their local congregations.

An instance concerning the office of the pope, otherwise known as the Bishop of Rome, underscores the authority of a sacred personage. The example, often referred to as the Avignon Papacy and the Western Schism, began with an incident in 1305.[37] In that year a Frenchman, Clement V, was elected pope. Under the influence of the King of France, Clement V moved the papacy from Rome in Italy to Avignon in France. The move led to the forfeiture of revenues from the Papal States in present-day Italy. In a bid to compensate for the lost income, Clement V charged fees for appointments, indulgences, and other services, provoking a wave of anti-clericalism across Europe.

The papacy remained in Avignon until 1378, at which time it was relocated back to Rome. That same year, the current pope, Pope Urban VI, sought to initiate a number of reforms, ending various corruptions and abuses within the Church. In particular, the pope sought to eliminate monetary rewards - gratuities, gifts, and benefices - for services rendered by Church officials. Unsurprisingly, conservative Cardinals, reluctant to forego the compensations to which they had become accustomed, opposed the reforms. In response, the Cardinals elected their own pope, Pope Clement VII, who established his papacy back in Avignon. Two popes now ruled, one from Rome and one from Avignon.

Royalty and commoners across Europe aligned their loyalties to their preferred pope. England, Denmark, and Norway were among the realms to follow Pope Urban VI. France, Naples, and Scotland, among others, favored Pope Clement VII. At this point, each pope excommunicated the rival pope and all the rival's followers. These actions excommunicated nearly all the people in Europe, meaning that, in principle (though not in practice), no one could confess their sins, and no one would reach salvation.

Attempts to resolve the situation, as much through coercion as from diplomacy, resulted in the election in 1409 of a third Pope, Alexander V, who possessed little power. Final reconciliation was achieved in 1417 with the election of Pope Martin V, who took office in Rome. The last of the antipopes, claimants descended from the Avignon Papacy, served from various locations until 1429 (some sources cite 1427), but with little support.

The Avignon Papacy and Western Schism touched the spiritual lives of nearly the entire population of Western Europe. It is, truth be told, an uncommon incident, yet it portrays the authority and breadth of influence of a sacred personage, as well as the interplay between Religion and mortal rulers. The schism diminished the Papacy's direct political power and, concurrently, strengthened the independence of England, France, and other nation states from ecclesiastical authority. Yet the Papacy survived, and its incumbents continue to guide the spiritual lives of over one billion followers worldwide.

The five sacreds constitute the major components of Religion's *how to think* process. Behind these components, however, lies a further catalog of techniques, each of which buttresses one or more of the five sacreds. The following list illustrates several of the more prosaic techniques.

MEDITATION Jainism, Bahá'í Faith, Sikhism, Hinduism, and many other religions employ the ritual practice of meditation. In its non-secular form, meditation is a spiritual exercise that allows the practitioner to demonstrate devotion to a faith, contemplate doctrinal teachings, or progress on a path to salvation. Buddhism, for example, emphasizes meditation as a means of attaining nirvana. Specific body postures and motions often accompany meditation, such as rhythmic breathing or sitting in a Lotus position. The repetition of holy words, the use of personal artifacts (amulets and prayer beads), and the selection of favored locations enhance the meditative experience. Preferred sites include temples and shrines.

CHANTING Chanting is the rhythmic singing of a mantra: a sacred sound, the enunciation of the name of a deity, or the recitation of verses from a sacred text. Similar to meditation, chanting offers a path to spiritual development. Through tone and iteration, chanting relaxes the body, focuses the mind, and facilitates contemplation of holy or metaphysical elements. In some religions, chanting promotes a direct connection with a deity. Many contemporary faiths embrace chanting, among them sundry denominations of Christianity, Buddhism, Hinduism, Islām, and Judaism. An explicit example is the Gregorian chant in Catholicism. Another is the Wiccan practice of *casting a circle,* a spell-like chant accomplished by means of spinning around multiple times while reciting sacred words aloud. Like meditation, chanting often is performed at sacred venues.

STUDY Study consists of formal and informal approaches. Clergy and scholars engage in formal study, which refers to the examination of the origins, deities, texts, traditions, and practices of a religion. Members of the laity often practice informal study, the less scholarly, periodic reference to sacred texts. Studying consecrated texts provides comfort, instills a sense of spirituality, and acts as an affirmation of faith. For those seeking guidance, the verses and tenets offer instruction on leading one's life in accordance with the faith.

CONSULTATION Consultation refers to engagement with a member of clergy: a guru, minister, priest, imām, or rabbi. The follower enlists clergy for advice on, and resolution of, specific issues of faith. Clergy are deemed to possess a greater understanding of the religion, its beliefs, and its traditions. In some faiths, such as Catholicism, clergy hear confessions and absolve sins, a practice that cleanses the soul of the penitent and engenders reconciliation with the deity. In religions without clergy, trusted lay members sometimes discharge similar responsibilities.

PRAYER Prayers, which represent an attempt to engage directly with a deity, either individually or communally, fall into three broad categories: supplication, thanks, and praise. Supplication, attempting to motivate a deity to perform (or prevent) a specific action, perhaps is the most common type. In religions with a personal god, the deity is deemed to listen to all prayers but does not necessarily grant all petitions. In common with meditation, prayer sometimes entails ritualistic body postures, such as genuflecting, bowing the head, placing the palms of the hands together with fingers pointed upward, touching the forehead to a carpet, orienting the body toward a shrine, or sitting cross-legged on a raised platform. One example from Judaism, common among supplicants at the Wailing Wall in Jerusalem, is swaying - bending forward from the waist, returning to an upright position, then repeating the cycle once per second. Here again, prayer rituals frequently prescribe the holding or wearing of religiously significant objects as, for example, a cross, a rosary, or a tallit. The act of prayer, like meditation and chanting, sometimes induces physiological effects. Relaxed breathing, a reduced heart rate, and similar corporeal responses not only soothe the practitioners but augment their belief in the efficacy of prayer.

INTUITION Religion presents intuition as a way of knowing, a means of absorbing celestial truth without conscious reasoning or empirical justification. I use the term *intuition* loosely, as the religious definition itself is fluid. The overlap with related vocabulary - instinct, inspiration, consciousness, emotion, premonition, feelings, common sense, and inner reasoning - is high. And the overlap varies by faith, context, and interpretation. Some religions posit intuition as the vehicle through which spiritual truth is conveyed. Others teach that a deity deliberately gifted intuition to humans for this purpose. Scripture buttresses distinct perspectives. Some Christian theologians, for example, interpret Biblical passages as an affirmation of intuition, as in "All scripture is given by inspiration of God."[38] Others equate intuition with references to the heart, where transcendent knowledge is felt and embraced: "When wisdom entereth into thine heart, and knowledge is pleasant unto thy soul."[39] Imāms offer similar interpretations: "Faith is knowledge in the heart."[40] Many Eastern religions present intuition as a path toward peace and enlightenment. In Jñāna yoga, a Hindu philosophy, it is intuitive wisdom that reveals the real self and produces liberation from suffering.[41] The Bhagavad-Gītā, a Sanskrit text that influenced both Hinduism and Buddhism, promises attainment of absolute truth through concentration of the mind on the deity Krishna, and by no other means.[42] Zen Buddhists till a singular furrow, mediating upon koans to exhaust the intellect, thereby opening the mind to intuitive insights. To cite a final example, sundry sects of Hinduism and Jainism teach that chakras (energy centers) inhabit the human body, performing both physical and spiritual roles. The former generates a life force while the latter acts as a psychic center of consciousness, a gifted locus adept at channeling metaphysical enlightenment.[43] Among these and still other interpretations, a common trait emerges: Intuition becomes a path to spiritual truth.

The preceding techniques, routinely practiced in nearly all religions, have proved highly popular. Billions of followers engage in one or more of the practices every day. One central tenet, however, stands supreme. It underlies the five sacreds and all of the associated techniques that comprise Religion's *how to think* process: authority. The authority of tri-omni deities, the authority of scripture, the authority of sacred persons, the authority of sacred places, and the authority of sacred rituals. Truth, morality, life, one's place in the universe - all derive from authority. The committed follower accepts and abides by this authority.

In this narrative, I chose to depict *how to think* as a process that precedes *what to think*. Yet the latter often precedes the former. The teaching of religious doctrine often begins with *what to think* at a young age, then is reinforced by *how to think* throughout adolescence and adulthood. As such, the five sacreds become a vehicle for reinforcing both *how to think* and *what to think*, creating a continuous, self-reinforcing cycle.

The entwinement of *how to think* and *what to think* produces Religion's canon of knowledge. From this canon, Religion provides answers to myriad key questions in life. What is the universe and why does it exist? Are we alone in the cosmos? What is the origin of life as a whole and of humanity in particular? Is there a supreme being? Are morals absolute or relative? What is my place in the world? Is there life after death? Your faith's answers to these questions constitute what you believe. My faith's answers constitute what I believe.

Depending on the belief system, the canon might respond as follows. Yes, there is a supreme being. Morality derives from the supreme being. There are absolute rights and wrongs. Treating your neighbor fairly is an absolute right. Stealing is an absolute wrong. Monogamy is an absolute right. Homosexuality is an absolute wrong. The supreme being knows you and sometimes responds to your requests for assistance. You will be rewarded for observing the rights and avoiding the wrongs, either in this life or in a future state of being. Paradise awaits a temporal passing.

Adherents of a specific faith adopt common answers to all such questions, as they derive from the same authority. Within religions founded on the premise of eternal truth, the responses remain constant and immutable.

Our discussion thus far might portray Religion solely as a product of ideology, distinct and apart from other aspects of society. Nothing could be more untrue. Religion has occupied a pre-eminent role throughout history, often serving as the central institution that imparted meaning to life.[44] In subsistence and war-prone times, Religion frequently offered dignity and hope to lives that the English philosopher Thomas Hobbes (1588-1679) described as "solitary, poor, nasty, brutish and short."[45] In society after society, over hundreds and thousands of years, Religion often has provided at least a modicum of care to society's helpless at a time when kings and governments offered none.

In the present day, Religion's temporal beneficence is manifest: the building of hospitals, the establishment of schools, the easing of poverty, and the espousing of ethical stands against corruption. Equally important, Religion strives to fulfill the psychological needs of humanity by promising hope, inspiration, and eternal paradise. It continues to be tightly interwoven into nearly all dimensions of society, forming the driving current in numerous nations and countless communities. Such societies embrace the wisdom and traditions of their faiths, reflecting them in their institutions and politics, their jurisprudence and architecture, and their music, art, and literature. For many followers, in truth, there is no distinction between religion and everyday life. Religion and society are inseparable. While some societies, particularly Western secular nations, strive to separate church from state, the distinction might be unknown in other communities.

Let's close this chapter by summarizing several elements. We established a start date and definition for organized Religion as, respectively, 100,000 BCE and belief in a supernatural entity. We explained how Religion arose as a means to explain the natural world and to offer protection from that world. At the time, there was no other plausible explanation for the events that affected people's lives. We then outlined humanity's transition to an agrarian society, a revolution that triggered - amid manifold other societal changes - a consolidation of religious beliefs.

Today, numerous religions and countless deities continue to exist throughout the world, representing sundry forms of animism, polytheism, monotheism, and other *-isms*. In nearly all such *-isms*, the five sacreds form the basis for *how to think* and *what to think*, each affirmed by nothing less than the underlying authority of a deity. It is this last point, as future chapters will detail, that is most critical to our narrative.

MISCELLANY

I use the term *hominin* to refer to the extinct and extant species of bipedal humans and their immediate ancestors, those who arose circa six million years ago. More precisely, this includes the genera *Homo*, *Australopithecus*, *Paranthropus* and *Ardipithecus*. The definition of hominin has evolved under various taxonomical re-classifications. Biology currently positions our species in a family, subfamily, tribe, genus, and species structure. The family *Hominidae* (aka hominid) includes the four great apes: gorillas, orangutans, chimpanzees, and humans. The subfamily *Homininae* excludes orangutans. The tribe *Hominini* excludes gorillas. The genus *Homo* excludes chimpanzees, leaving humans. We belong to the species *Homo sapiens*, the only extant human species. Some sources suggest *H. sapiens* emerged some 300,000 to 400,000 years ago, considerably earlier than the 200,000 years cited in the main text.

Scholars seeking to uncover the first written record of unambiguous religious activity often turn to Mesopotamia and Egypt. Mesopotamia is the historical region bounded by the Euphrates and Tigris rivers, encompassing most of present-day Iraq and parts of several surrounding countries. One of the earliest Mesopotamian civilizations was Sumer, first settled around 5000 BCE. Sumerian scribes etched texts onto clay tablets using a wedge-shaped script known as cuneiform. Many of the tablets, which date from 2600 BCE, document secular topics: inventories of goods, receipts, laws, and poetry. Others, however, record the deities, prayers, and rituals of Sumerian society. Egyptologists, in similar form, point to the Pyramid Texts, which are thought to date from roughly the same period. The texts, carved on stone walls and sarcophagi, attempt to safeguard pharaohs and other prominent individuals in their celestial afterlife.

For narrative simplicity I consolidate the time spans of human activity into three broad eras: Paleolithic, Neolithic and modern. A more accurate rendering of eras and ages appears in the *Postscript*.

Humanity's transition from hunter-gatherer to agrarian societies is considerably more intricate than outlined. Diamond provides an excellent treatment of the subject, for which he earned a Pulitzer Prize. He suggests that the Neolithic Revolution began in the Fertile Crescent, a region of temperate weather where people had access to a wide variety of domesticable plants and animals.

509 BCE marks the foundation of the Roman Republic, which was displaced by the Roman Empire in 27 BCE. 476 CE marks the end of the Western Roman Empire, by which time Rome had been sacked twice: in 410 CE by the Visigoths and in 455 CE by the Vandals.

Some sects of Judaism and Islām do not see Christianity as monotheistic, owing to the Christian concept of the Trinity. The Trinity is the belief that God appears as three distinct personages - as Father, Son, and Holy Ghost - yet with a single common essence. The belief arose in the first century CE. Christian theologians debated the meaning, roles, and hierarchy of the essences for three hundred years before settling the issue at the councils of Nicea and Constantinople in the fourth century. Christians, of course, consider their religion monotheistic.

The primary translation methodologies of sacred and non-sacred texts are formal equivalence (word-for-word) and dynamic equivalence (thought-for-thought). Proponents of the former advocate a *this is what it says* translation. Exponents of the latter espouse a *this is what it means* translation. Most translations reflect a weighted mix of the two approaches.

Pythia - or, as often expressed, *the* Pythia - was the name of the High Priestess at Delphi. Many individuals, of course, occupied the role as High Priestess, who dispensed prophecies for well over a millennium.

A summary of the definition, start date, and *how to think* process of Religion. Diagram 2.6

Summary of Religion	
Definition	Belief in a supernatural being
Start Date	Circa 100,000 BCE
How To Think Process	The Five Sacreds
How To Think Core Tenet	Authority

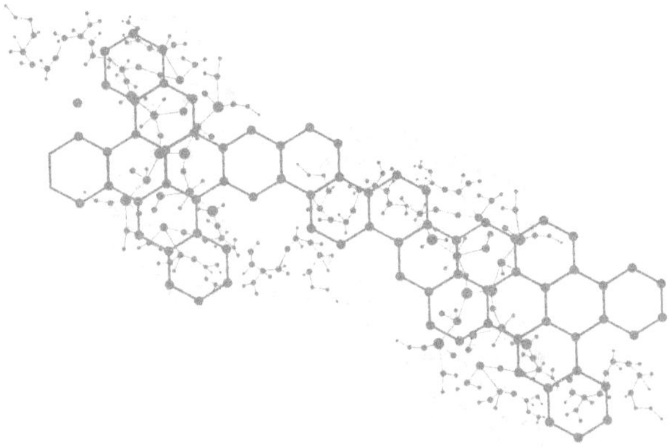

[III]

The Origin Of Reason

To surrender to ignorance and call it God has always been premature, and it remains premature today.[46]

ISSAC ASIMOV (1920–1992) AMERICAN AUTHOR

We embarked on our examination of Religion by establishing a definition and a start date. Let's do the same with Reason, building a definition from three of the common branches of reasoning: deductive, inductive, and abductive.

The first type, deductive reasoning, is a form of logic championed by Aristotle (384-322 BCE),[47] in which the truth of a conclusion follows from its premises. Deductive reasoning is a general-to-specific form of reasoning, where a set of general premises leads to a specific conclusion. Diagram 3.1 provides classical and not-so-classical examples.

In each case, if the premises are true, then the conclusion is true. It should be evident that establishing the truth of the premises is paramount. The second example is logically correct in that conclusion follows argument, but one should not rush to purchase a neck guard.

The second type of reasoning, inductive reasoning, is predicated on causation and observation (empiricism). In direct contrast to deductive reasoning, induction is a specific-to-general form of reasoning. Induction begins with observations within specific areas, then formulates a general conclusion from the observations. Diagram 3.2 provides an example from the British Empiricist philoso-

Diagram 3.1
Examples of deductive reasoning.

	Classical	Non-Classical
Premise 1	All men are mortal.	Vampires exist.
Premise 2	Socrates is a man.	Vampires drink human blood.
Conclusion	Socrates is mortal.	Buy a neck guard.

Diagram 3.2
Hume's archetype of inductive reasoning.

Premise	The Sun has risen in the east every morning.
Conclusion	The Sun also will rise in the east tomorrow..

pher David Hume (1711-1776), extracted from his masterwork *A Treatise of Human Nature*.[48]

It is important to note that, even though the premise might be true, inductive reasoning does not guarantee the validity of the conclusion. Simply because the sun has risen in the east 1,000 times does not ensure it will rise in the east the 1,001st time. So why is inductive reasoning relevant? There are three reasons. First, as noted, it is based on observation rather than abstract logic. Second, it allows prediction of as-yet unobserved phenomena - in this case, that tomorrow the sun will rise in the east. And third, inductive reasoning acts as an unparalleled mechanism for confirming correct predictions and disproving incorrect ones. As we will see, all three aspects are central to our narrative.

The American logician Charles Pierce (1839-1914) is credited as one of the first to define the third type of reasoning, abductive reasoning. He laconically differentiated abductive and inductive reasoning as, respectively, facts seeking a hypothesis and a hypothesis seeking facts.[49] Abductive reasoning favors the best explanation among competing hypotheses, either by disproving rival accounts or by demonstrating that one explanation best fits observation. Abductive reasoning is a commonplace in medicine where an illness might have several causes, with the physician selecting the most likely diagnosis.

The scientific method, which we explore at length later in the chapter, is based heavily on inductive reasoning. So much so that I often use induction and the scientific method as synonyms, though it is not formally correct to do so. We will see that abductive reasoning also features in the scientific method. Deductive reasoning, however, can be a bit of a mixed bag. It occupies a strong position in logic and mathematics, where proofs rely on deductive arguments, each founded on axioms and inferences, to demonstrate the truth of their assertions. Yet elsewhere, it often plays a smaller role due to its reliance, in its purest form, on rules of inference as the principal source and test of knowledge.

A synthesis of the above produces the definition of Reason that we will use: *the objective search for knowledge* informed by observation and testable predictions. One further point before we leave this topic. The level of overlap among the definitions of Reason, science, and the scientific method is high. Rather than cultivating confusion with three, similar-sounding denotations, however, we will use *the objective search for knowledge* for all three terms. Here are other apposite definitions:

[Science is] a series of judgments, revised without ceasing.[50]

ÉMILE DUCLAUX (1840–1904) FRENCH MICROBIOLOGIST

[Science is] an imaginative adventure of the mind seeking truth in a world of mystery.[51]

SIR CYRIL HERMAN HINSHELWOOD (1897–1967) ENGLISH CHEMIST

Science is the great antidote to the poison of enthusiasm and superstition.[52]

ADAM SMITH (1723–1790) SCOTTISH PHILOSOPHER

Science is a way of thinking much more than it is a body of knowledge.[53]

CARL SAGAN (1934–1996) AMERICAN ASTROPHYSICIST

With definition in hand, we shift our attention to antiquity. What of the origin of Reason? When, where, and how did it arise? Who were its first proponents? To unearth answers to these inquiries, we return once again to the Paleolithic.

The journey began with early, intelligent pioneers who took the first steps towards shaping a rational view of the world. Perhaps Bokn, a cousin of Zaug, was among the pioneers, observing the moon from night to night. Perhaps Bokn notched marks on a bone, counting the number of nights from one full moon to the next, calculating the cycle to be 29 nights. Perhaps he organized his notches into groups, each cluster reflecting a phase of the moon, concluding that the earth's satellite cycles through eight principal phases, as depicted in Diagram 3.3. Equally inquisitive individuals at locations around the world recorded similar observations, noting the patterns of the moon, the location of the sun rising and sinking on the horizon, the changing levels of water in a lake, or the yield of crops grown in various soils. This was an ad-hoc, sporadic process, often based on faulty analysis and yielding incorrect conclusions. But it was a start.

As Paleolithic transitioned to Neolithic, the same transformations that set the stage for organized Religion also erected the framework for Reason. The expansion of food production, the consequent release of labor from farming, the growth in populations, and the rise of cities laid the foundation for the first proto-scientific professions. These transformations, coupled with extended periods of societal stability, provided early practitioners with the time and means to practice their endeavors. Enlightened leaders acted as another catalyst, of which more shortly.

The early to mid-Neolithic Period witnessed varied practical advances, among them refinements in animal husbandry; breakthroughs in metallurgy; insights on soil mechanics;[54] and improvements in civil engineering for earthworks, megalithic structures, and deep mining. Importantly, this era also recorded the invention of the wheel and the development of writing. This last invention - the encoding of language onto physical materials that permit the retrieval and distribution of knowledge - stands out. It accelerated the acquisition and dissemination of knowledge in the late Neolithic, by which time the faint light of Reason had begun to glimmer in sundry locations: from the Middle East to India, and from China to Mesoamerica. Perhaps the best-documented case, however, lies at the tip of southeastern Europe, in ancient Greece.

Grecian societies of the first millennium BCE not only reached the necessary level of development for Reason to prosper, but likewise handed down extensive written records. We refer here not to the country of Greece but to the roughly five-hundred to one-thousand (estimates vary) Greek city-states and their environs. Among

Diagram 3.3
The eight principle phases of the moon.

these, Argos, Athens, and Thebes were notable poleis. The height of Greek culture occurred in the aptly-named Classical Age, around the fifth to third centuries BCE. A fair number of eminent individuals from the era are still widely recognized today. Herein a sampling.

The Mathematicians Pythagoras (c.570-c.495 BCE), Aristarchus (c.310-c.230 BCE), Eratosthenes (c.276-c.195 BCE), and Archimedes (c.287-c.212 BCE).

The Philosophers Xenophanes (c.570-c.475 BCE), Anaxagoras (c.510-c.428 BCE), Zeno (c.490-c.430 BCE), Socrates (470-399 BCE), Plato (c.428-c.348 BCE), and Aristotle (384-322 BCE).

The Physicians Hippocrates (c.460-c.370 BCE) and Galen (129-c.216 CE).

The Playwrights Aeschylus (c.525-c.456 BCE), Sophocles (c.495-c.406 BCE), Euripides (c.480-c.406 BCE), and Aristophanes (c.446-c.386 BCE).

The above and their peers, many of them polymaths, contributed to diverse fields of knowledge: geometry and astronomy, biology and medicine, philosophy and history, architecture and sculpture, and democracy and literature. From the quantity, diversity, and sagacity of Greek contributions, scholars often point to the Classical Age as the cradle of Western civilization - an epitaph both admirable and enviable for a culture of relatively limited population. It might be surprising to learn, however, that the Greeks are not considered the first scientists, as their objectives and techniques often differed from current scientific practice. An apt label for the Greeks and other early pioneers is *natural philosopher*, a commonly-used term before the modern age of science. This nomenclature does not diminish the collective contribution of the Greeks but merely frames it in context of the age in which it occurred.

Reason, from its flowering in ancient Greece, trekked westward with the rise of the Roman Empire. The Romans are not widely recognized as substantial contributors to Reason, but notable exceptions exist. One is Pliny the Elder (23 CE-79), the author and natural philosopher who compiled a 37-volume encyclopedia reflecting the knowledge of the era.[55] Another is Cicero (106-43 BCE), the statesman, orator, and philosopher whose corpus continued to influence humanist and republican principles 1,500 years after his death.[56] Let's also accord to the Romans plaudits for assimilating Greek ideas and innovations, thereby safeguarding Hellenic science, philosophy, and arts for succeeding generations.

After its Roman sojourn, Reason reversed course and turned eastward, journeying to the Islāmic world. A proximate cause of the migration was the decline of the Western Roman Empire. Sources differ on the start date of the decline, but it no doubt had begun by the year 455 with the sack of Rome at the hands of the Vandals, culminating in the year 476 with the fall of the Emperor Romulus Augustus (461-c.507). The ebb ushered in the suitably-named Middle Ages in the West, an interval of 1,000 years throughout which Reason and science progressed slowly, inconsistently, or not at all. Reason was frequently muzzled or suppressed when its rational worldview clashed with the vested interests of religious authority. As we will see in *The Case of Gravity*, the Catholic Church, the dominant religious institution in medieval Europe, routinely exercised its authority to repress "deviant" thoughts and behaviors. But as the West dropped the baton of Reason, the East was prepared to grasp it.

The Arab world's embrace of Reason began with a mass translation movement that began in the eighth century and continued for 200 years.[57] Arabic scholars sought and acquired Greek, Persian, Indian, and other texts from throughout the Middle East and beyond. The various writings inscribed on paper, papyrus, and parchment were translated into Arabic, providing Islām with a deeper base of knowledge and a broader worldview than that attainable in any other civilization of the era.

What led to the translation movement? One factor was the inspiration of enlightened leaders. The Abbāsid caliphate, which founded the city of Baghdād, produced several such leaders.[58] The most prominent was Abū Ja'far Abdullah al-Ma'mūn (786-833), who initiated both the translation movement and the greatest period of scholarship since ancient Greece. This period became celebrated as the Golden Age of Islām, paralleling the reign of the Abbāsids from 770 to 1258. It was accompanied by the Mu'tazili school of Islāmist theology, which prevailed from the eighth to the tenth century. The theology, some scholars suggest, promoted a liberal interpretation of reason and inquiry as consistent with Islām: "Seeking knowledge is a duty upon every Muslim."[59]

Al-Ma'mūn appears to have extended academic tolerance to both Islāmic and non-Islāmic scholars. Baghdād, then other cities, attracted mathematicians, engineers, physicists, chemists, astrologers, philosophers, and scholars of all stripes. This critical mass of scholarship, once attained, became self-sustaining. The presence of Greek-speaking Christians, who had knowledge of Greek science and who now lived in lands governed by Islām, further bolstered the translation movement. Possibly more far-reaching, however, was the arrival of paper. The acquisition of paper-making technology from China rendered it simpler and cheaper to record and disseminate knowledge.[60] Indeed, it has been suggested that ninth-century Baghdād represented the first time in human history when it was possible to make a living from writing and selling books in the marketplace.

The Golden Age of Islām produced notable scholars from many lands (Afghan, Arab, Berber, Egyptian, Persian) and from many religions (Christian, Jewish, Muslim, Zoroastrian), a quantity far too numerous to enumerate. Here are a few of the eminent thinkers, often unknown in the West, that instantiate the era:

Ibn al-Haytham Abū Ali al-Hassan ibn al-Hassan ibn al-Haytham (c.965-1039), Arab physicist; employed elements of the scientific method; contributed to mathematics and astronomy; among his extensive works, authored the *Book of Optics*.

Al-Bīrūni Abū Rayhān Muhammad ibn Ahmed al-Bīrūni (973-1048), Persian polymath; contributed to mathematics, astronomy, history, and linguistics; measured the size of the earth to higher accuracy than previously; regarded as the father of geodesy.[61]

Al-Kindi Abū Yūsuf Ya'qūb ibn Ishāq ibn al-Sabbāh al-Kindi (c.801-c.873), Arab philosopher and polymath; contributed in mathematics, physics, medicine, and other fields; translated texts from Greek to Arabic; referred to as the philosopher of the Arabs.[62]

Ibn Sīna Abū Ali al-Hussein ibn Abdullah ibn Sīna (980-1037), Persian physicist, philosopher, and physician; author of the *Canon of Medicine*; arguably the most eminent scholar in Islām.[63]

It is important to acknowledge the contributions of Indian and Chinese scholars in this narrative. From 320 to 550 CE India experienced its own age of discovery, appropriately named the Golden Age of India. It was a period of relative peace and prosperity, with advances in physics, mathematics, and the humanities. The works of many Indian scholars were translated into Arabic during the Golden Age of Islām. One example is the *Āryabhatīya,* a treatise on astronomy and mathematics written by the pandit Āryabhata (467-550). Among the topics covered in the tract, Āryabhata calculated the length of the solar year, approximated the value of π, and described place value notation in the decimal number system. The *Zhou Bi Suan Jing,* a text from around 200 CE (or possibly earlier), illustrates similar contributions from China. The text delineated solutions to 246 mathematical problems, including an early proof of the Pythagorean theorem.[64]

The translation movement reached an end late in the tenth century. By that time, most of the great scholarly works had been located and translated. The Golden Age itself came to an end two centuries later. The causes are debated.[65] Perhaps it was the invasion of the Mongols and the destruction of Baghdād in 1258, which forced the Abbāsids to relocate to Egypt. Perhaps it was the rise of a more literal rendition of Islām, asserting that the revealed truths of the Qur'ān and the Hadīth would provide all the wisdom required by Muslims. Regardless of cause, the East would soon pass the baton of Reason back to the West, a baton capably managed for five centuries. And the West, at last, was primed to receive it. The catalyst was the Renaissance.

The Renaissance in the West began in the fourteenth century in northern Italy, spreading gradually across Europe. The term *Renaissance* is a French word, itself derived from Latin, for *rebirth*, a *mot juste* marking the rediscovery and reemergence of learning. The Renaissance was a broad cultural movement, first affecting European intellectual discourse, then everyday life in science and philosophy, religion and politics, and literature and art.

Knowledge reintroduced to Europe from the Muslim world often entered through Spain, where Islām continued to exercise influence until the fall of Granada in the fifteenth century.[66] Along with routes through Italy and elsewhere, Europe embraced and began to enlarge the ancient learning of the Greeks, Persians, Arabs, Indians, and Chinese. The printing press with movable type, invented in the fifteenth century in Germany by Johannes Gutenberg (c.1398-1468), marked the foremost innovation of the age. The invention proliferated rapidly, initiating the mass production of pamphlets and books, and swelling the circulation of ideas and information in a few short decades.

Following on the heels of the Renaissance came the Enlightenment, a flowering of scholarship that emerged in the seventeenth century. In one sense similar to the Renaissance, the Enlightenment was a philosophical and intellectual movement. But unlike its predecessor, the Enlightenment fundamentally reshaped Western thought by emphasizing Reason as the primary source of knowledge and truth. This period, also aptly known as the Age of Reason, initiated a new worldview that elevated empirical evidence and scientific inquiry above dogmatic platitudes.

Central to the Enlightenment was belief in the power of human intellect. Philosophers like John Locke, Immanuel Kant, and Voltaire not only advocated for the use of Reason, but rejected the traditional doctrines put forward by church and monarchy. This shift encouraged the questioning of established beliefs and the pursuit of knowledge through observation and experimentation.

The importance of the Enlightenment to Reason and science is profound. It laid the groundwork for the modern scientific method, an inductive approach where general principles are derived from specific observations, and where the same principles form the basis of reasoned investigation. Four of the many individuals who championed the power of Reason and the use of mathematical principles in explaining natural phenomena appear below.

Copernicus Nicolaus Copernicus (1473-1543), Polish mathematician and astronomer; formulated a mathematical model of the universe that placed the Sun rather than Earth at its center; launched the Copernican Revolution, a paradigm shift in the understanding of the cosmos.

Bacon Francis Bacon (1561-1626), English statesman and natural philosopher; often referred to as the father of empiricism;[67] published his seminal work, *Novum Organum Scientiarum* (which Google translates as *The New Instrument of Science*) in 1620, laying out principles of the scientific method that influenced subsequent generations of scholars; notably rejected syllogism in favor of induction.[68]

Galileo Galileo Galilei (1564-1642), Italian astronomer and physicist; routinely challenged dogmatic authority; posited that the laws of physics are the same for all objects moving at a constant speed, regardless of speed or direction; the first modern thinker to assert that the laws of nature are mathematical and to confirm their operation via inductive reasoning.

Newton Isaac Newton (1643-1727), English polymath whose work exemplified the power of Reason; articulated the laws of motion and universal gravitation; measured the masses of planets and predicted the motions of comets; calculated the oblateness of the Earth and described the propagation of sound; arguably the greatest scientist in human history.

In parallel to the accomplishments of ingenious individuals, the Enlightenment also witnessed the establishment of scientific societies and institutions, such as the Royal Society in England and the *Académie des sciences* in France. These organizations facilitated the exchange of ideas and the dissemination of scientific knowledge. Publications like Denis Diderot's *Encyclopédie* explained and spread Enlightenment principles, making scientific knowledge more accessible to a broader audience.

This framework of education and intellectual exchange spawned countless advancements across all spheres of human activity, from agronomy to chemistry, from astronomy to psychology, and from biology to medicine. Each discovery, from its foundation on Reason, not only undermined superstition and religious canon, but paved the way for secularism and the separation of science from religion.

In summary, the Enlightenment was crucial to the development of modern thought. It fostered an environment that valued critical thinking and empirical evidence, leading to the establishment of principles that continue to underpin scientific inquiry today. Indeed, inquisitive societies around the globe continue embrace Reason, in whole or in part, extending the groundwork laid by their inquisitive ancestors.

This completes our brisk examination of Reason's journey from its initial Paleolithic origins to its first solid landfall in ancient Greece, its migration to the East, then its revival in the West and beyond.

One task remains before we proceed. We have yet to establish the start date for Reason. Some sources align the start date to the publication of a notable text that ***a)*** employs elements of the scientific method, and ***b)*** articulates a consequential scientific discovery. Archetypal works include Copernicus's *On the Revolutions of the Celestial Spheres*, published in 1543, and Galileo's *Dialogue Concerning the Two Chief World Systems*, published in 1632. Others sources cite the establishment of constitutional monarchy in Britain (1689) or the onset of the French Revolution (1789). Still others suggest a later date, correlating the start of Reason to the genesis of science as a profession (rather than an amateur undertaking) or to the espousal of a non-religious worldview.

I choose 1543 for the start of Reason, the earliest date cited above. Three considerations drove the selection. First, Copernicus' opus marked an abrupt departure from the prevailing geocentric model of the universe, one that held sway for 1500 years. Second, it directly challenged religious dogma as a valid source of knowledge. And third, *On the Revolutions* exemplifies a handful of tomes that, collectively, helped to instill a new mindset, one that helped launched the scientific revolution.

How does this date compare to that of Religion? Recall we established 100,000 BCE as the start date of religious belief. Reason, by contrast, emerged a mere five centuries ago. This chronology of origins explains, in large part, the profound influence of religious dogma. It dominated human thought and action for well over ninety-nine percent of the history of our species. In many parts of today's world, it continues to do so. Reason's *how to think* process, by comparison, is a relative infant. Its youth, however, belies its uniqueness, its clarity, and its potency - traits to be explored in this and upcoming chapters.

THE SCIENTIFIC METHOD

We now turn to the principal *how to think* process of Reason, the scientific method, components of which were developed independently in separate parts of the world. In the West, the seed of the method was planted in the sixteenth century. The cultural openness of the Renaissance permitted it to germinate, while the subsequent era, the Enlightenment, allowed it to flower. The method is depicted as the 5-step process shown in Diagram 3.4. The depiction is unconventional but will help to articulate a number of salient points. The remainder of this chapter outlines the steps of the method, while subsequent chapters explore its application.

STEP 1 - ASK

The scientific method starts by asking a question, a question regarding a phenomenon that one seeks to understand. It is likely one has contemplated the phenomenon, conducted some informal observations, or explored prior research on the topic. From this initial study, the inquirer might conclude that an understanding of the phenomenon is incorrect, incomplete, or non-existent. Or perhaps the curious individual simply wishes to reconfirm an established principle. In all instances, further inquiry is required.

As the first crucial step toward the advancement of knowledge, Ask requires that inquiries be framed clearly and objectively. Sound inquiries employ clear definitions, cite explicit premises, and permit alternative outcomes. Unsound queries, by contrast, embrace ambiguity, bias, and false dilemmas - to name but three of many specious practices.

One notably essential trait warrants special mention: openness. Both the individual and the society must permit open inquiry. On a personal level, the inquirer must be open-minded and willing to accept unexpected results. He or she must avoid inquiries explicitly designed to ratify existing beliefs. On a societal level, the cultural environment must allow questions to be asked, without subjecting the questioner to ostracism, legal punishment, or threat of violence. As we will see, at many points in history the contemporary culture

Diagram 3.4
The scientific method, the principal *how to think* process of Reason.

precluded inquiry, particularly on topics that challenged orthodox beliefs.

To illustrate Ask and the subsequent steps of the scientific method, we will reference Galileo and his examination of falling bodies. What was Galileo's inquiry? He wanted to know whether bodies of identical material but unequal mass fall at the same or different rates. Aristotle (384-322 BCE) taught that objects fall at different speeds in relation to their mass. The heavier the object, the faster it falls.[69] Aristotle, however, often reached conclusions from abstract logic rather than from experiment. Galileo sought an evidence-based conclusion.

Before leaving Ask, let's note that subjective qualities figure no less prominently than objective criteria in driving inquiry. Examples of the former include speculation, creativity, and beauty - attributes which frequently inspire inquiry. They enable an individual to envision the world through a singular lens, deviate from mainstream thinking, and propose what initially might be considered outlandish ideas. A researcher possessing these qualities examines data and evidence, yet pursues insights that depart from the norm, often risking ridicule from peers. Such insights frequently lead to dead-ends, but occasionally drive consequential discoveries and precipitate new paradigms. To illustrate, the English physicist Paul Dirac (1902-1984) sought beauty in equations as predictions of reality.[70] This approach led Dirac to develop an aesthetically elegant equation that explains how certain subatomic particles behave when they travel close to the speed of light. The equation was the first to combine elements of quantum mechanics with Einstein's special theory of relativity. Moreover, it predicted a new form of matter, antimatter. Dirac, whose speculations subsequently were confirmed by experiment, was awarded the Nobel Prize in 1933. Not bad for beauty.

STEP 2 - PREDICT

The second step begins with the formulation of a hypothesis. A hypothesis is a suggested explanation derived from a small set of possibly correlated or causal data pertaining to explicit scenarios. A hypothesis is a precursor to a theory, which follows later in the process. The principal attribute of a hypothesis is that it must be testable through observation and experiment. When tested, a hypothesis will be accepted or rejected. Making a hypothesis testable requires predictions. If the hypothesis does not yield testable predictions, the hypothesis is invalid.

The English polymath William Whewell (1794-1866) established three criteria for the testability of a hypothesis. The hypothesis should predict novel phenomena, not merely elucidate existing phenomena. The hypothesis should explain a variety of phenomena, not solely a narrow range. And the hypothesis should exhibit coherence - aligning to a broader scientific theory or a comprehensive body of knowledge.[71]

In the twentieth century, the Austrian-British philosopher Karl Popper (1902-1994) articulated a fourth requirement. Each hypothesis must be falsifiable.[72] That is, there must be an experimental result or future discovery that could disprove the hypothesis. Let's make sure we understand this concept. Consider this hypothesis. The blue whale is the largest mammal on Earth. Could this hypothesis be falsified, that is, shown to be false? Yes, simply find and measure a more enormous mammal. Note that falsifiable means *capable of being disproved*. It does not mean *will be disproved*. The blue whale, by the way, continues to retain its crown. To date, no larger mammal has been identified.

It is impossible to overstate the importance of falsifiability. No quantity of positive outcomes can prove a hypothesis, but one negative outcome can disprove it. If a hypothesis cannot be invalidated, the hypothesis is unsound, subsequent experiments will be unsound, and conclusions drawn from the experiments will be unsound. If I claim, for example, that an invisible witch cast a spell on a redwood tree, causing it to mushroom to a remarkable height, the claim is not falsifiable. No test can refute the claim. Hence the claim itself is not credible.

As we will see, the ability of testable and falsifiable predictions to confirm or refute hypotheses is unparalleled. No other philosophy, no competing ideology, no metaphysical doctrine possesses a comparable tool. Not one. As such, testable and falsifiable predictions comprise the core tenet of the entire scientific method. Underlying this tenet is the trait of skepticism, which Batson cites as one of the three unique features of the scientific method (along with empiricism and systematic research). In this context, skepticism "leads the scientist to construct a situation in which the [hypothesis] can show its falseness." If a hypothesis, he continues, "is a good one, a fairly explicit statement of the falsifying conditions can be made."[73]

It should be noted that skepticism does not imply robotic disbelief. Instead, a skeptic begins with a sort of *tabula rasa*, a blank slate on the topic under consideration. The skeptic then seeks a reasonable body of evidence as the basis for reaching reliable and valid conclusions. The skeptic eschews comforting, convenient, diplomatic, and impulsive judgments. Equally, skeptics value the method by which a conclusion is reached more highly than any particular outcome.

Framing testable and falsifiable predictions requires care. Pitfalls abound. To cite one instance, one must avoid the allure of the "same fact twice."[74] A hypothesis cannot be concocted on known evidence, with predictions validated from the same evidence. To demonstrate, suppose, upon entering a building, you notice a vehicle in the parking lot with the license plate number AXG 279. Once inside the building, you construct a hypothesis that there is a vehicle in the parking lot with the plate number AXG 279. You then return to the parking lot, conduct an observation, and record a result confirming your hypothesis. It should be evident the hypothesis is invalid, as it was framed to be self-fulfilling.[75]

The "same fact twice" gimmick demonstrates another pertinent aspect of Predict. The framework for testing a hypothesis must be developed before data is gathered. Methodologies for collecting and analyzing observations must be planned before experiments begin. That is, the methodology must clarify the specific outcomes that will either confirm or refute the hypothesis. Failing to do so risks a biased selection of results engineered to produce a desired outcome. As Elsberry states, standard scientific procedure annuls "the deprecated practice of examining a data set after the fact and then figuring out what statistical test will produce a significant result of the desired form."[76]

Returning to our example from Step 1, Galileo may have constructed the following hypothesis. Two spherical objects of identical material but unequal mass fall at the same speed. This hypothesis offers a testable prediction. Simply drop some objects and see what ensues. The hypothesis offers differentiation of outcomes. Either the objects hit the ground at the same time, or they don't. The hypothesis also is falsifiable. If the two balls strike the ground at different times, the hypothesis is refuted. As an aside, Galileo's actual hypothesis was slightly more complex, as it related to bodies of unequal mass falling with uniform acceleration. But we will continue with the simpler version.

STEP 3 - OBSERVE

It is time to put the hypothesis to the test. This entails executing the previously-designed experiments to ascertain the validity of the hypothesis. The nature of the experiments will vary according to the hypothesis being tested, but all rely on empiricism, which is the principal attribute of this step in the scientific method. Empiricism, as articulated by the British philosopher John Locke (1632-1704) and others, refers to reliance on observation (empirical evidence) in the determination of experimental outcomes.[77]

Initially, empiricism was built upon observations by human senses, but today is more commonly based on measurements by scientific instruments. This is an important distinction. Sense-based observations often are less accurate, particularly when there is a delay between making and recording an observation. The delay opens the door to the inaccuracies of human memory. Human biases likewise enter the picture. Favoritism and prejudice, whether veiled or overt, unwitting or calculated, distort accurate measurement. Indeed, eyewitness, earwitness, and experiential testimony may be prominent within courts of law, but they are the lowest forms of observational evidence in science.[78] Precision instruments and robust procedures drive the collection of sound observational data.

This step of the process also includes analysis of the experimental results, for which many comparative and mathematical methods exist. The rigorous, skeptical, and dispassionate evaluation of the evidence leads to a decisive juncture: the moment when the hypothesis is accepted or rejected. To quote Richard Feynman (1918-1988), one of the greatest physicists of the second half of the twentieth century, "It does not make any difference how smart you are, who made the guess, or what his name is - if it disagrees with experiment it is wrong."[79] Period. This reality affirms that a hypothesis always must be tentative, and it must be subject to abandonment if disproved. Note further that deductive reasoning plays little to no role in Observe. The evidence of empirical data quashes the most insightfully constructed, astutely reasoned, and exquisitely articulated arguments.

To test his hypothesis, Galileo rolled balls down an inclined plane, measuring their speeds and arrival times at regular intervals. Multiple iterations of this experiment allowed Galileo to test and accept his hypothesis. Two objects of identical material but unequal mass do fall at the same speed.

STEP 4 - VERIFY

Verify refers to the elevation of a broad set of confirmed hypotheses into an accepted theory. Let's pause a moment and define the word *theory*. In everyday speech, its meaning can border on the trivial. "I was sitting on the balcony last night, enjoying a glass of wine while watching the coded blinking of fireflies darting among the trees. More wine and more fireflies led me to the theory that the twinkling stars in the cosmos might, in reality, be distant, celestial fireflies seeking to attract mates." This is a thought, a hunch, an idea, a reflection, a speculation, perhaps even a meditation, but it is not a theory. In our context, a theory has a more robust, scientific definition. A theory is a well-substantiated explanation of a broad range of phenomena, founded on a comprehensive body of evidence, offering testable and falsifiable predictions. The word *comprehensive* is all-important, as many ideas (and alleged theories) are proposed and defended on minuscule data, selected observations, or subsets of evidence. Critically, a theory does not become a theory by virtue of the position, intelligence, reputation, prestige, or strength of personality of its proponent.

As stated, a theory offers a broader and more compelling set of testable predictions than a hypothesis. And to be accepted, a theory must be independently verified. Independent verification demands a theory's predictions be tested by someone other than the author of the theory. Ideally, this requires a separate set of experiments performed by separate individuals, preferably in separate laboratories and locations. This practice reduces the risk that evidence, favorable or unfavorable, issues from chance, bias, error, inexperience, ineptitude, or deception.

Attestation by someone who holds a strong stance against a theory ofttimes instills even greater credibility in the theory. The American physicist and Nobel Prize winner, Robert Milliken (1868-1953), for example, reportedly expended considerable energy attempting to disprove Einstein's particle theory of light, which Milliken believed was incorrect. In 1915, after ten years designing and conducting experiments to disprove the theory, he publicly (and perhaps reluctantly) confirmed Einstein's predictions.[80]

In our simple example, we can now say Galileo's hypothesis has been elevated to a theory, independently verified multiple times over the centuries. One such verification was performed by the Flemish physicist Simon Stevin (1548-1620), which he documented in the treatise *The Practice of Weighing* (*De Weeghdaet*)[81] in 1586. Stevin's experiment was notable for two reasons. First, whereas the story that Galileo dropped balls from the Leaning Tower of Pisa is apocryphal, Stevin actually performed the trial. He dropped two lead balls, one ten times heavier than the other, from the Nieuwe Kirk (a church) in the Netherlands. He confirmed both balls struck the ground at the same instant, concurrently validating Galileo's hypothesis and stating flatly "Aristotle is wrong."[82] Second, unknown to the Italian physicist, Stevin published his validation 52 years earlier.

STEP 5 - ACCEPT

At some point, a valid theory will have passed a range of testable predictions. It will have explained all the observations of reality that preceded it, and it will have correctly predicted the results of novel phenomena. The weight of evidence then compels acceptance of the theory, and it is deemed to be true. Having passed the rigors of the scientific method, the theory now joins the body of knowledge of Reason. It is in this manner that Reason employs its *how to think* process to build its *what to think* canon - slowly, meticulously, inexorably, with each discovery forming the foundation for the next insight.

To close our example, Galileo's theory on the speed of falling bodies has been accepted for nearly four centuries. But that is not quite the end of the process, as there are two important caveats. The first arises when a theory runs afoul of tenaciously-held secular or religious beliefs. This can spawn a desire to dismiss or reject the theory, despite a preponderance of substantiating evidence. The second caveat is that acceptance of a theory does not guarantee its overarching validity. New evidence at a future date could challenge the theory. This is a regular and critical component of Reason. Science is a self-correcting process (though not always in the short-term) as new evidence comes to light. Perhaps for this reason, accepted theories are no longer commonly referred to as laws, as no law is correct 100% of the time. We will revisit both caveats in future chapters, but before proceeding, let's expand a bit more on the potential impact of new evidence. One example concerns gravity.

Newton provided a comprehensive description of gravity in his masterwork, *Mathematical Principles of Natural Philosophy*, published in 1687. He based his theories of motion and gravitation on the concepts of absolute space and absolute time.[83] In this context, the word *absolute* means that space and time exist as independent entities, each unaffected by the other. Simply put, his theory of gravity states that every object containing mass attracts every other object. The more massive the objects, the greater the attraction. The greater the distance between the objects, the weaker the attraction.[84] His theory provided many testable predictions on the orbits of planets and the speeds of falling objects. So widely accepted was his theory that his views dominated much of cosmological thinking for the next three centuries (and in many instances, still do). Two cases illustrate the dominance and evolution of Newton's theory.

In the first half of the nineteenth century, astronomers noticed irregularities in the orbit of the outermost planet of the Solar system, Uranus, 2.6 billion kilometers distant from Earth. The irregularities caused the orbit of the planet to differ from Newton's predictions. In 1845 the French mathematician Urbain Le Verrier (1811-1877) and the British astronomer John Adams (1819-1892) independently predicted the presence of a new planet beyond Uranus, one whose mass would account for the irregularities.[85] Their predictions relied on the correctness of Newton's theory. With what result? In 1846 the German astronomer Johann Galle (1812-1910) located the hidden planet, which we now call Neptune. The work of Le Verrier, Adams, and Galle yields a splendid example of prediction confirmed by independent observation.

In the second half of the nineteenth century, the situation repeated itself. In possession of more accurate observational data, it was observed that something was amiss with the orbit of the planet Mercury. Its movements did not quite follow the predictions of Newton's theory, but this time there was no new planet. A different explanation was required. Enter Albert Einstein.

Einstein (1978-1955) was born in Germany and migrated to the USA in 1933, where he remained for the rest of his life. His theories of special relativity (1905) and general relativity (1915) yielded a new description of gravity. Instead of Newton's absolute space, Einstein suggested that space distorts in the presence of massive objects - it curves.[86] Diagram 3.5 depicts this concept.[87] Einstein further combined Newton's absolute space and absolute time into a single, non-absolute spacetime, proposing that both space and time change in the presence of mass.

In Newton's world, a more massive object (say, the sun) attracts a less massive object (say, the earth), and a bit of vice versa. In Einstein's world, the more massive sun causes an indentation in spacetime, and the less massive earth moves along the resulting curve,

Diagram 3.5
The curvature of spacetime as described by Albert Einstein.

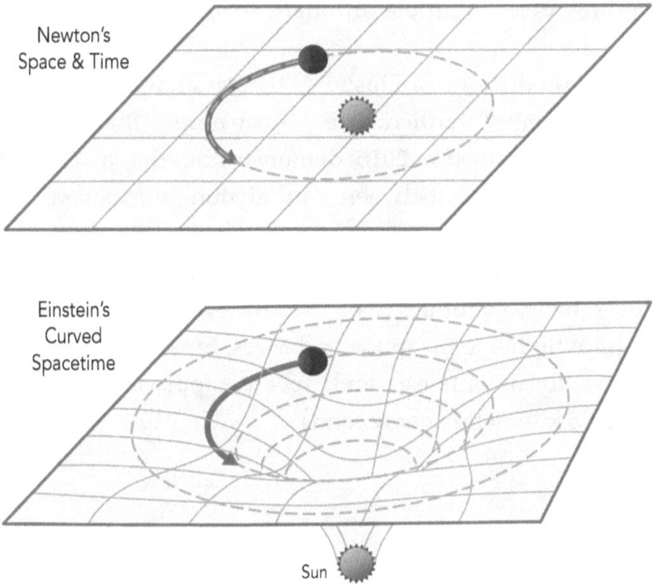

tracing what we perceive as its elliptical orbit. Without the curvature of spacetime, the earth would travel in a straight line. Did Einstein's theories of gravity explain Mercury's orbit? Yes, perfectly.[88]

Einstein's theories illustrate the role of testable predictions in three ways. First, the theories fully explained existing (Newtonian) observations of gravity. Second, they explained a new phenomenon, the anomaly of Mercury's orbit, which the previous theory was unable to explain. And third, they introduced a new set of testable predictions. To name several: Einstein's theories predicted that time slows (clocks run more slowly) as speed increases, light bends in the presence of gravity,[89] and gravitational waves (caused by massive, accelerating objects) ripple through spacetime at the speed of light.[90]

All the predictions of Einstein's theories have been repeatedly and independently verified. It is worth noting that gravitational waves were confirmed in 2016, demonstrating that a lengthy interval sometimes elapses between a prediction and its confirmation. One conclusion that should not be drawn from this example is that Newton's theory of gravity was incorrect. It simply was incomplete. Newton's theory continues to be used for everyday tasks like calculating space flights to the moon and Mars. Einstein's theories apply where gravity is intense, and speeds approach that of light (299,792,458 kilometers per second).

For brevity and clarity, I presented the scientific method as a clean and sequential process inevitably leading to progress. The reality differs. It is iterative, and it often is slow and messy. Hypotheses are poorly constructed. Assumptions prove to be invalid. Mistakes occur. Samples are mislabeled. Experimental results are misinterpreted. Ingrained philosophies hinder advances. Prevailing societal beliefs distort impartiality. Egos outstrip objectivity. Dishonesty trumps integrity. Disputes arise. Tempers flare. Controversy ensues. Theorists pursue false paths for decades. Years pass without notable

Diagram 3.6
The bending of starlight, a prediction of Albert Einstein's general theory of relativity, verified in 1919.

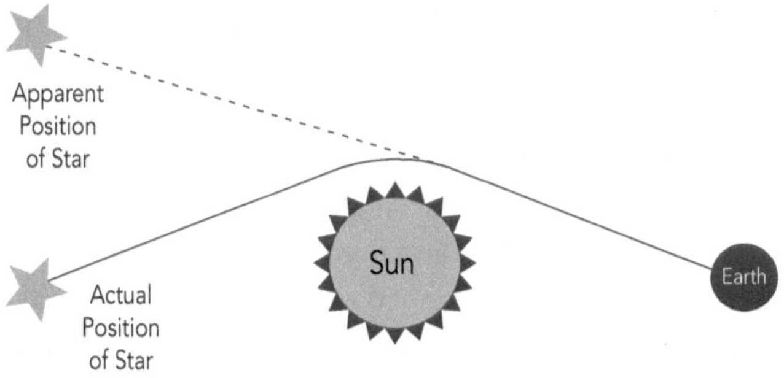

results. Researchers publish findings in slivers to lengthen their bibliographies. Journals express disinterest in negative findings. Academia's publish-or-perish mindset cultivates piffle. One laboratory is unable to reproduce the results of another. Reported findings omit crucial detail. Funding is withdrawn. And, as the American physicist Thomas Kuhn (1922-1996) noted, science periodically undergoes revolutions, creating new paradigms, new approaches, and new understandings that might not have been previously considered.[91]

There also exist multiple variations of the scientific method. Each field of inquiry possesses singular characteristics which require alterations in approach. Some require simple statistical probability techniques to explain cause-and-effect relationships; others require number theory and matrix algebra. Some support the study of repeatable phenomena; others require a comparative or historical approach. Some allow a small number of variables to be isolated and controlled; others face dauntingly complex interdependencies. Some enable researchers to touch and manipulate the object of interest; others require techniques that operate at unfathomable distances and scales. All demand creativity, intelligence, deliberateness, close attention to detail, and occasional serendipity.

The most meticulously planned and rigorously executed experiments still can yield surprises. The humble mouse offers a contemporary example. Laboratory mice are carefully bred for genetic uniformity at tightly-controlled facilities, then purchased by research institutions around the world. Among the sundry murine strains bred with distinct traits, the most common is the Black Six, a mouse used for research in medicine, genetics, psychology, and other disciplines. It has long been known that variations in the handling and care of mice can affect the outcome of clinical trials. Known variations include food, light, temperature, and humidity. That said, recent studies suggest that additional factors may play a role: trials on littermates versus trials on mice living apart; trials on mice handled by men versus mice handled by women; and variations in the microbiome (the bacteria that live on and in the mice) among otherwise identical mice.[92] Such findings, if verified, call for even more stringent planning and execution of experiments, and in some cases raise questions on the validity of past results.

Regardless of approach and experimental subtleties, however, all valid scientific experiments share a key trait. They encompass the single most critical attribute of the scientific method: testable and falsifiable predictions.

Here again, we close a chapter by summarizing its key elements. We cited humanity's transition to agrarian societies and the associated stability of those societies as prerequisites for Reason. We identified the scientific method as the principal *how to think* process of Reason. The process relies on open inquiry, well-formulated hypotheses, empirical evidence, independent verification, and a willingness to accept unpopular conclusions. It is testable and falsifiable predictions, however, that form the core tenet of the entire process. Together, they offer an unequaled ability to debunk false or poorly constructed hypotheses.

We also affirmed that the consistent and rigorous application of the scientific method produces Reason's *what to think* body of knowledge. And not just any assortment of *what to think* miscellany. It yields a sound, cogent, and defendable *what to think* canon. As such, the ideas that survive Reason's methodology stand on firmer ground than those proclaimed by pseudoscientific and dogmatic belief systems. To quote physicist and author Jim Al-Khalili (1962-), "The modern scientific method is *non-negotiable* in its all-encompassing importance as a worldview. Many would argue it is the *only* worldview that a rational, thinking person can have in explaining how and why the world is the way it is."[93] This is precisely the claim of Reason.

MISCELLANY

Sherlock Holmes, the fictional detective created by Sir Arthur Conan Doyle (1859-1930), was famous for his powers of deduction. The opening chapter of *The Sign of Four* (published in 1890) even is entitled *The Science of Deduction*. But most of his reasoning was inductive.

This text positions Greece as the scholarly seedbed of Western civilization. Yet the accomplishments of the venerable Athenians, Ionians, and Thebens were not entirely of their own invention. Earlier cultures - among them the Egyptians, Akkadians, Phoenicians, and Achaemenians - contributed to the writing systems, mathematical foundations, religious beliefs, and agricultural practices of the ancient Greeks. Scholars recognize these and similar knowledge transfers, but debate their timing and extent.

The origin of the Pythagorean theorem is disputed. Some suggest it first was discovered in Babylon. Other scholars point to Egypt, India, China, and elsewhere, where a more complete understanding of the theorem arguably emerged. Early Greek and Roman literature attribute the discovery to Pythagoras (c.570-c.495 BCE), the Greek philosopher and mathematician, as the eponymous source of the theorem.

Who was the first scientist? There is no agreed answer to this question. Some point to Ibn al-Haytham, Francis Bacon, or Galileo Galilei, each of whom employed experiment, inductive reasoning, or other elements of the scientific method. Others suggest Nicolas Copernicus or Issac Newton, citing their publication of consequential, paradigm-shifting texts. This author favors the Greek philosopher Miletus of Thales (c.626-c.545 BCE). Although many of his assertions proved to be incorrect, he is thought to be the first individual to seek natural rather than supernatural explanations of the world.

The term "scientific method" is an accurate appellation, in that the methodology arose from various fields of science. To some, however, the moniker implies an artificial constraint. It suggests the methodology is valid only within the natural sciences (e.g., physics, chemistry, and the earth sciences). The methodology, in fact, is valid across all spheres of human endeavor - in the social sciences, in politics, and beyond. Perhaps one day it will be called the "universal method".

A summary of the definition, start date, and *how to think* process of Reason. Diagram 3.7

Summary of Reason

Definition	The objective search for knowledge
Start Date	1543 CE
How To Think Process	The Scientific Method
How To Think Core Tenet	Testable Predictions

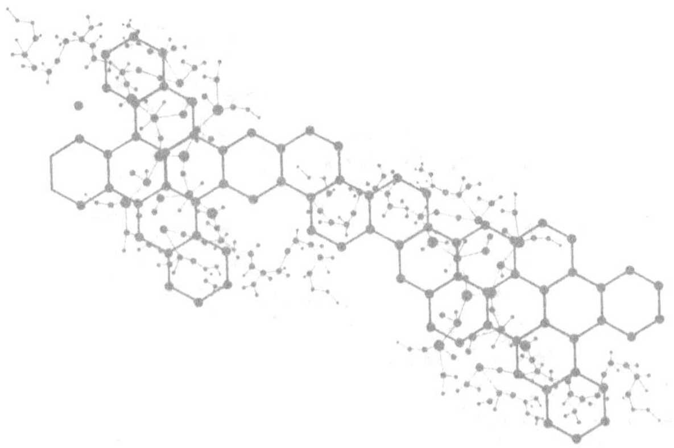

[IV]

The Cases

Science, at bottom, is really anti-intellectual. It always distrusts pure reason, and demands the production of objective fact.[94]

H. L. MENCKEN (1880–1956) AMERICAN JOURNALIST

We have completed our synopses of Religion and Reason. We outlined their origins, chronicled their evolutions from Paleolithic to present day, and described their principal *how to think* processes - the five sacreds and the scientific method. We saw that authority underlies the five sacreds: the authority of sacred deities, sacred texts, sacred places, sacred rituals, and sacred persons. Similarly, we identified empiricism and testable predictions as the underpinnings of the scientific method.

We now examine three cases where the world acquired new knowledge, one case each from the seventeenth, eighteenth, and nineteenth centuries. In each instance, Religion employed the full vigor and authority of its five sacreds to determine its *what to think* conclusions. Reason followed a parallel approach, enlisting the capability and credibility of its scientific method. The three cases share a common element. In each instance, the respective *how to think* processes of Religion and Reason produced disparate *what to think* outcomes. This dichotomy allows us to probe the *how to think* process of each domain, assessing its ability to generate valid conclusions. The same dichotomy illustrates how each sphere reacts to the *what to think* outcomes produced by the opposing domain.

To illuminate salient points in the narrative, each case emphasizes selected steps of the scientific method as described in the following paragraphs. In the real world, of course, Reason rigorously employed all five steps of the method to reach its conclusions.

The first case is *The Case of Gravity,* in which we ascertain whether the force of gravity that causes an object to fall toward the surface of the earth is the same force that holds the moon in its orbit. To reach a determination, we begin with the work of Galileo (1564-1642) and his *Dialogues Concerning the Two New Sciences,* published in the seventeenth century. We then proceed through a ten-step exercise, applying only those techniques accessible to natural philosophers from the ancient era to the early Enlightenment. The result allows us to examine the opposing views of Religion and Reason. Gravity itself was not a point of contention between the two domains, but its effects were divisive. Do stars orbit in perfect circles or ellipses? Do planets travel at uniform or variable speeds? Are the celestial heavens eternal and unchanging? More fundamentally, why did Religion elect to adopt theological views on these topics? *The Case of Gravity* investigates these subjects. The case also highlights the following steps of the scientific method.

ASK Inquiry is the first step in the search for new knowledge. How does each domain react to inquiry? What happens when authorities construe inquiry as a threat to orthodox teachings and beliefs? What befalls the inquirers?

PREDICT The ten-step solution depicted in *The Case of Gravity* punctuates the importance of detail. Detail instills confidence. Detail provides the foundation for formulating a hypothesis. Detail provides the basis for testable and falsifiable predictions, the key attribute of the entire scientific method. Without detail, there exist no grounds to authenticate or accept a claim. How do faith-based assertions address detail? What is the role of prediction in religious prophecy? What is the role of teapots?

How old is the earth? This is the question we explore in the second case, *The Case of Earth's Age*. Religion and Reason produce dramatically different answers to this question, their estimates differing by nothing less than billions of years. While Religion bases its estimates on scripture and revelation, Reason studies natural physical processes. Reason, arguably, began to cultivate its first serious estimates of the planet's age in the eighteenth century, based on the blossoming science of geology. The English physician James Hutton (1726-1797), often cited as the founder of modern geology[95], placed the discipline on a sound scientific footing. From observation and experiment, he argued that natural processes continually reform the surface of the earth over geological timescales. What estimates have Religion and Reason provided for the age of the earth? Which assessment is correct? How do *natural philosophers* differ from *natural theologians*? *The Case of Earth's Age* unravels these puzzles while illuminating two more steps from the scientific method.

> OBSERVE Reason requires systematic procedures and empirical evidence. What techniques does Reason employ to obtain a diverse set of evidence? Does Religion possess a process similar in form to Observe?

> VERIFY Reason demands independent verification of its theories. How does Reason utilize one experimental technique to corroborate the findings from a second technique? Which domain makes its data and methods transparent, allowing nonpartisan verification of its claims? How does Religion verify its claims?

The final case, *The Case of Human Origins*, distinguishes between the fact of evolution and the mechanism that propels it: the theory of natural selection. The seminal event in *The Case of Human Origins* was the publication in 1859 of *On the Origin of Species* by the scientist Charles Darwin (1809-1882). Reason offers Darwin's theory to explain the origin of humanity. Religion advances a contrasting explanation. Which one is more likely to be true? What evidence does each domain provide? *The Case of Human Origins* resolves this enigma and provides insights on the last step of the scientific method.

> ACCEPT Reason sets a high bar, requiring that a theory ***a)*** pass a range of testable predictions, ***b)*** explain all the observations of reality that preceded it, and ***c)*** correctly predict the behavior of novel phenomena. A preponderance of evidence then drives acceptance of the theory. What threshold does Religion apply for acceptance of its claims? What techniques does Religion employ when confronted with unwelcome evidence?

We refer above to Charles Darwin as a scientist. The term *scientist* was coined by William Whewell around 1840 to denote a "cultivator of science,"[96] one who undertakes a systematic investigation of the world. Whewell minted the word to differentiate the practitioners of philosophy from those of science. The distinction paints the former as thoughtful and judicious rationalists, while the latter roll up their sleeves to conduct observations and experiments - the work of science. Darwin was among the first to earn the newly-coined appellation. It was well-deserved.

Let's proceed to the three cases.

The cases and steps of the scientific method to be examined in the next three chapters.

Diagram 4.1

The Cases

Case	Scientific Method
The Case of Gravity	Ask
	Predict
The Case of Earth's Age	Observe
	Verify
The Case of Human Origins	Accept

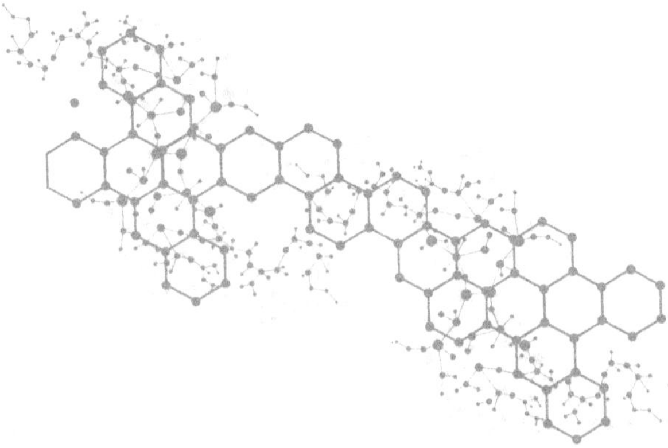

[V]

THE CASE OF GRAVITY

Extraordinary claims require extraordinary evidence.[97]

CARL SAGAN (1934–1996) AMERICAN PHYSICIST

We begin with *The Case of Gravity*. Galileo, the Italian physicist and astronomer whom we met in chapter three, will assist us. We quickly will revisit his inclined plane experiment, by which he was able to approximate the force of gravity. Using nothing but the knowledge available to natural philosophers before the year 1700, we will proceed through ten short steps to calculate the force of the earth's gravity pulling on the moon. Our goal is to determine whether the gravity that pulls an object toward the surface of the earth is the same as the force that holds the moon in its orbit. Our hypothesis, which we will confirm or reject in the tenth step, is that the same force of gravity applies in both instances. Diagram 5.1 outlines the ten-step approach.

Galileo's experiment will prove insightful by casting light on a number of topics. Why were gravitational effects divisive? How did Religion develop its dogmatic view of reality? One that asserted a geocentric universe. One that averred a starry firmament bathed in luminiferous aether. And one that tasked stars and planets, each embedded in one of eight eternal and changeless celestial spheres, to perpetually trace flawless orbital circles. How did Religion react when Reason suggested a different reality? What is the role of detail in credible explanations of phenomena? *The Case of Gravity* answers these questions and, in doing so, permits a closer examination of the first two steps of the scientific method, Ask and Predict.

This is the only chapter that contains mathematical calculations. I include the calculations to demonstrate that the numbers, in fact, do add up. If you prefer, read the text, review the figures and skip the arithmetic. Let me further remark that none of the figures is drawn to scale, each is merely descriptive. In some cases, I use modern measurements to improve the accuracy of the calculations. I explicitly state these instances in the text. Let's proceed to Step 1 on page 96.

Diagram 5.1
Ten steps to compare the force of Earth's gravity on the surface of the earth and its force on the moon.

Step	Description
①	Find the force of gravity on Earth.
②	Find the circumference of the earth.
③	Find the radius of the earth.
④	Find the diameter of the moon.
⑤	Find the distance to the moon.
⑥	Find the circumference of the orbit of the moon.
⑦	Find the speed of the moon's orbit.
⑧	Find the fall of the moon.
⑨	Apply the Inverse Square Law.
⑩	Assess the result.

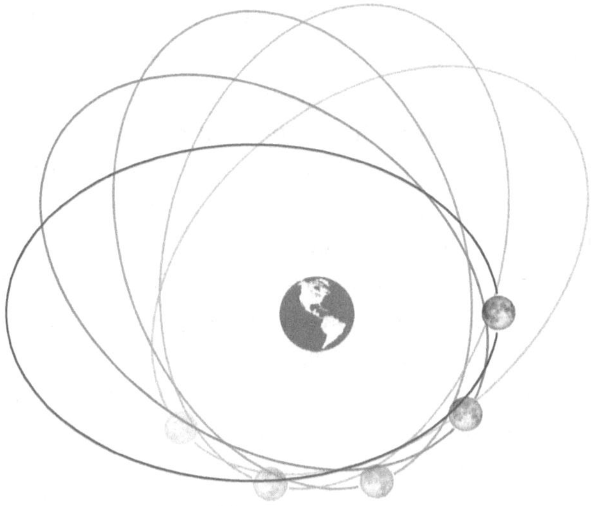

STEP 1
FIND THE FORCE OF GRAVITY ON EARTH

In 1638 Galilei published one of his greatest works, *Dialogues Concerning the Two New Sciences*, summarizing his work in physics. The two sciences refer to the science of materials and the science of falling objects. In the latter science, Galileo conducted experiments with falling objects to measure their rates of acceleration and, from their rates, the force of gravity.[98] Ideally, he would have released an object from a known height (say, 10 meters above the ground) and allowed it to fall straight downward, measuring its acceleration at various points over the course of its descent. In Galileo's time, however, no technology existed to measure the high speed of directly falling objects. He required a different method to observe and measure an object's fall. According to his writings, his solution was to construct an inclined plane (a ramp), and then to roll metal balls down the slope. The angle of the slope, and the consequent slower travel of the balls, allowed Galileo to record his measurements.

The following sentences provide a simple description of Galileo's experiment. As depicted in Diagram 5.2, he positioned bells at set lengths above the plane: at the ¼-unit, ½-unit, ¾-unit, and 1-unit points. He then released a bronze ball from the top of the ramp, timing the duration of its travel until it struck the first bell at the ¼-unit mark. He was able to measure the speed using a water clock synchronized to his pulse. He then removed the first bell and repeated the experiment, timing the duration of the ball's travel until it struck the second bell at the ½-unit mark. He repeated this process for the bells at the ¾-unit and 1-unit points, conducting the experiment multiple times until he was confident of the various travel times of the ball.

After completing his experiment, Galileo analyzed his data. He was able to confirm that the ball traveled to the ¼-unit point in half the time that it took to travel to the 1-unit point. By rolling balls of different weights, Galileo also was able to determine that all bodies fall with the same acceleration. He likewise showed that the distance traveled is a function of the square of the time (the ball will travel a greater distance in each successive time interval). Put another way, in one second the ball travels one unit of distance, in two seconds the ball travels four units of distance, and in three seconds nine units of distance. Notably, Galileo's results were in direct contrast to Aristotle, who believed the ball would travel the same distance in each successive time interval. This finding demonstrates the superiority of observation-based evidence over abstract logic.

Galileo's experiment also revealed the force of gravity. His measurements showed that, upon release, a ball experiences a fall of 488 centimeters in the first second. The 488-centimeter fall of the ball is a measure of the force of gravity on the surface of the earth. Let's remember this number. Write it down if necessary. We will return to it in Step 10.

STEP 1

A depiction of Galileo's inclined plane experiment to determine the force of gravity on Earth.

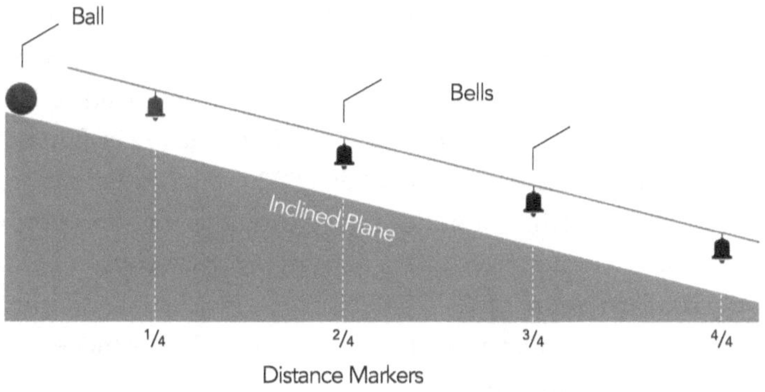

Diagram 5.2

STEP 2
FIND THE CIRCUMFERENCE OF THE EARTH

There are various methods for estimating the circumference of the earth. Al-Bīrūni, for example, derived his estimate from the heights of mountains. We, however, will use the method attributed to the Greek mathematician Eratosthenes (276-194 BCE), who served as the chief librarian at the Library of Alexandria. Eratosthenes knew that at the summer solstice with the sun directly overhead, the sun's rays reached all the way to the bottom of a well in Syene in southern Egypt. On the same date and at the same time of day, Eratosthenes measured the angle of the shadow created by a stick placed vertically in the ground in Alexandria in northern Egypt. He calculated the angle as 7.2°. He then measured the distance from Alexandria to Syene, which was determined to be 5,000 stades. Geometry then permitted Eratosthenes to calculate the circumference of the earth.[99]

The stade is an archaic measure equating to either 157 meters (the Egyptian stade) or 185 meters (the Olympic stade). It is not known which stade Eratosthenes used. I use the lower figure in my example, which yields a value of 39,250 kilometers for the circumference of the earth. But either figure provides a good approximation of the circumference, with a margin of error between 2% to 15% of the modern figure of 40,075 kilometers. Incidental and unrelated to our narrative on gravity, the techniques of Eratosthenes and Al-Bīrūni both demonstrated the earth was spherical, not a flat disc.

STEP 2

Applying the method of Eratosthenes to measure the circumference of the earth.

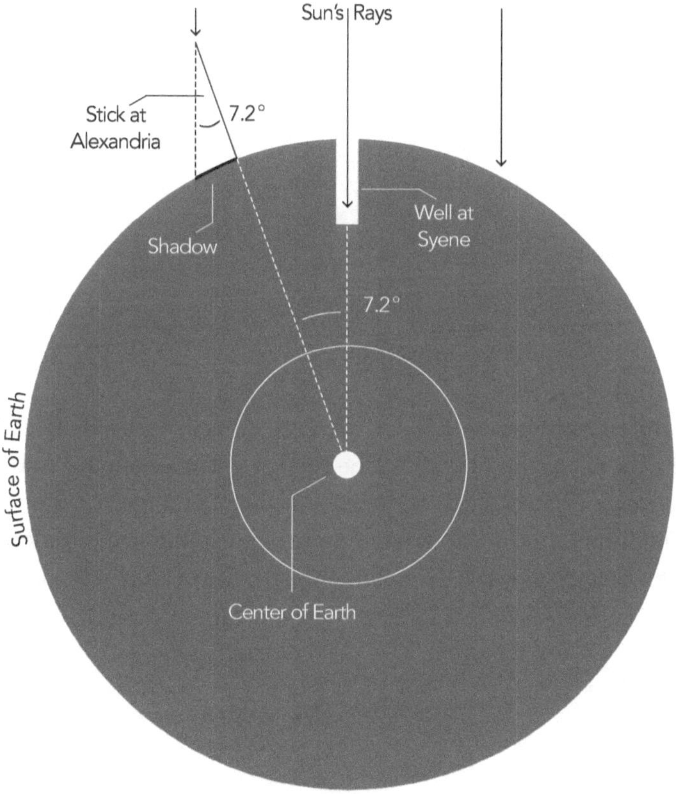

THE MATH

Circumference of Earth = (360° / 7.2°) * 5,000 stades
= 250,000 stades * 157 meters per stade
= 39,250 km

Diagram 5.3

STEP 3
FIND THE RADIUS OF THE EARTH

From the circumference of the earth, Eratosthenes was able to calculate the radius of the planet via the further use of geometry and the well-known relationships of circles. Using the modern value of 40,075 kilometers for the circumference, but following the method of Eratosthenes, yields 6,400 kilometers for the radius of the earth.

STEP 3

Using geometry to calculate the radius of the earth.

THE MATH

$$\text{Radius of Earth} = \text{circumference} / 2\pi$$
$$= 40{,}075 \text{ km} / 6.28$$
$$= 6{,}400 \text{ km}$$

Diagram 5.4

STEP 4

FIND THE DIAMETER OF THE MOON

The next challenge is to determine the diameter of the moon, which can be accomplished by careful observation during a lunar eclipse. Lunar eclipses vary in type and duration, each lasting from one to four hours. We will assume a duration of three hours for our example. To estimate the diameter of the moon, one observes the movement of the moon during the eclipse and records two measurements: the time required for the moon to pass completely into the earth's shadow (45 minutes), and the time required for the moon to cross the earth's entire shadow (180 minutes). The ratio of these two measurements provides an estimate of the diameter of the moon relative to the diameter of the earth. Completing the calculation produces a value of 3,200 kilometers for the diameter of the moon.[100]

STEP 4
Measuring the diameter of the moon during a lunar eclipse.

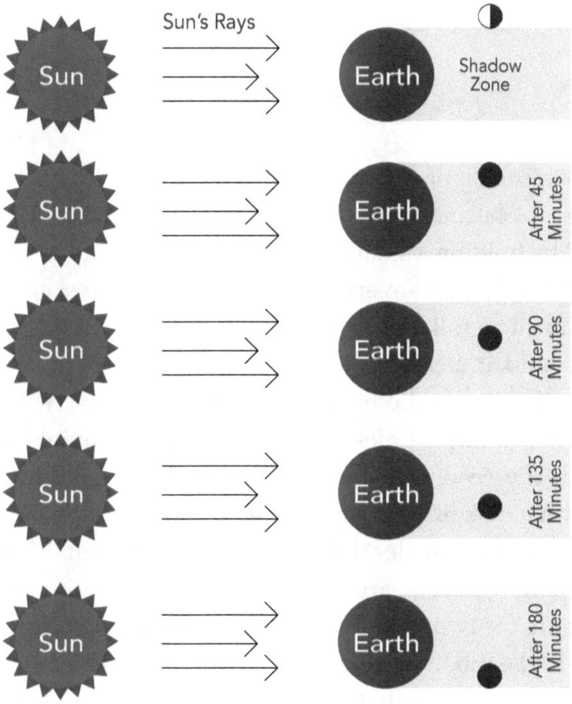

THE MATH

Diameter of Moon = (45 / 180) * diameter of Earth
= (45 / 180) * (2 * 6,400 km)
= 3,200 km

Diagram 5.5

STEP 5
FIND THE DISTANCE TO THE MOON

From his estimate of the diameter of the moon, Eratosthenes now was able to calculate its distance from the earth. This can be achieved by holding a small circular disk at arm's length, then positioning the disk to just overlay the moon. This is most easily accomplished during a full moon. One must find a disk of the right diameter and hold it steadily enough at arm's length (or use a fixed mount) to just cover the perimeter of the moon. My wife and I attempted this multiple times, and we can attest it requires some practice. With my 66-centimeter arm length, I was able to cover the moon with a disk 5.5 mm in diameter. My wife, with shorter arms, required a disk of smaller diameter. Try this yourself. It's fun.

The length of your arm and the diameter of your disk supply the measure for two sides of a triangle. One can now calculate the distance to the moon by comparing this triangle to a larger triangle reaching from the earth to the moon. This yields a distance of 384,000 kilometers.

STEP 5

Using similar triangles to measure the distance to the moon.

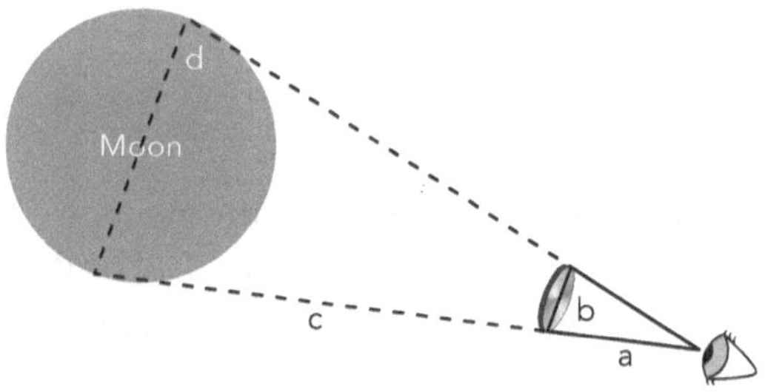

THE MATH

a = Distance from eye to disk
b = Diameter of disk
c = Distance to moon
d = Diameter of moon

$$c / d = a / b$$
$$c = (a / b) * d$$
$$= (66 \text{ cm} / 5.5 \text{ mm}) * 3{,}200 \text{ km}$$
$$= 384{,}000 \text{ km}$$

Diagram 5.6

STEP 6

FIND THE CIRCUMFERENCE OF THE ORBIT OF THE MOON

To calculate the circumference of the orbit of the moon, we can use the formula $2\pi r$ to find the circumference of a circle. The distance to the moon calculated in Step 5 provides an estimate of the radius of its orbit (the distance from the earth to the moon, and the radius of the moon's orbit, are the same number). The calculation yields a value of 2,411,250 kilometers for the circumference of the moon's orbit.

STEP 6

Using geometry to calculate the circumference of the moon's orbit.

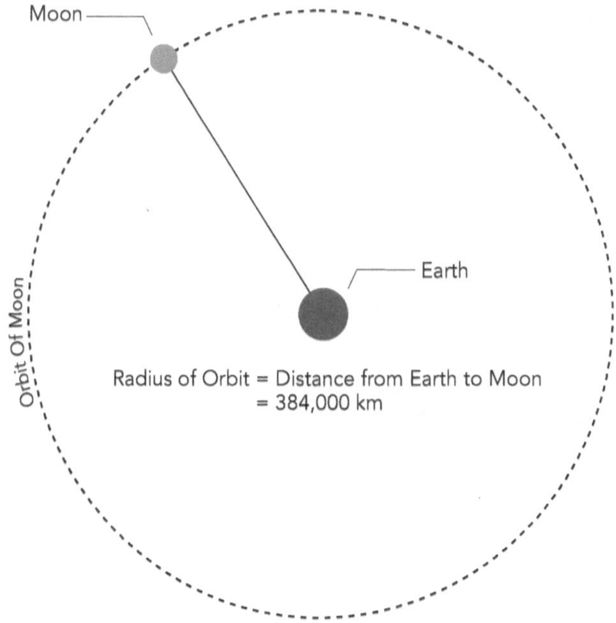

THE MATH

$$\text{Circumference of orbit} = 2\pi * \text{distance to moon}$$
$$= 2\pi * 384{,}000 \text{ km}$$
$$= 2{,}411{,}520 \text{ km}$$

Diagram 5.7

STEP 7

FIND THE SPEED OF THE MOON'S ORBIT

Ancient civilizations were able to calculate the lunar month based on observation, with estimates ranging from 27 to 30 days. We will adopt the modern revolution period of 27.3 days, then calculate the speed of the moon's orbit by dividing the circumference of its orbit by its period. The calculation indicates the moon travels 1 kilometer per second.

STEP 7

Calculating the speed of the moon's orbit.

THE MATH

Speed of orbit = 2,411,520 km / 27.3 days
= 88,334 km per day
= 1 km per second

Diagram 5.8

STEP 8

FIND THE FALL OF THE MOON

Without gravity, the moon would travel in a straight line. Gravity causes the moon to *fall* toward the earth, and to adopt a slightly elliptical orbit. The eccentricity of the moon's orbit, however, is small (0.055) and does not affect our calculation. Eccentricity is the measure of deviation from a perfect circle. A perfect, circular orbit would have an eccentricity of zero.

The Pythagorean theorem provides a method for calculating the fall of the moon by imagining a right triangle extending from the orbit of the moon to the center of the earth. The triangle will be much skinnier than depicted. The adjacent side (***a***) of the triangle is the distance to the moon, calculated in Step 5. The opposite side (***b***) is the speed of the moon's orbit, calculated as kilometers per second in Step 6. The Pythagorean theorem ($a^2 + b^2 = c^2$) yields the length of the hypotenuse (***c***) of the right triangle. Subtracting the length of the adjacent side from the length of the hypotenuse provides the fall of the moon, 0.0000013 kilometers in one second. The modern value is 0.00000136 kilometers in one second, which is equivalent to 0.136 centimeters in one second.

STEP 8

Using geometry to calculate the fall of the moon.

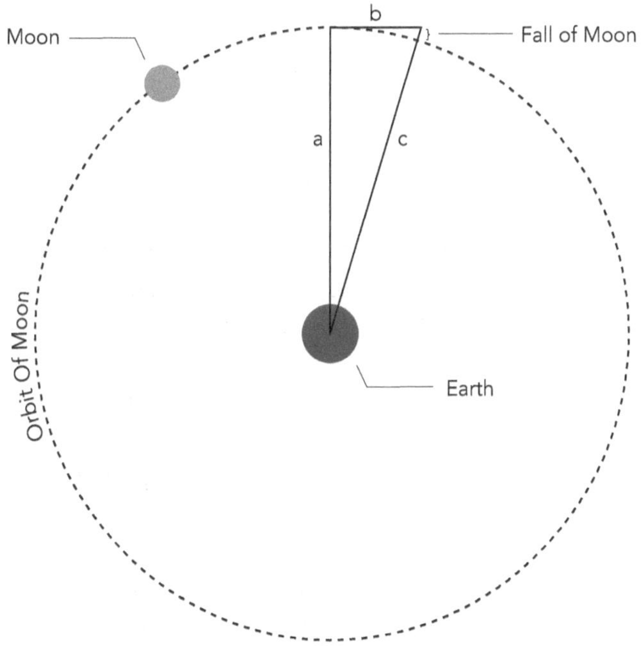

THE MATH

a = distance from center of Earth to orbit of Moon
b = speed of Moon's orbit per second
c = length of hypotenuse of right triangle

$$\begin{aligned}\text{Fall of Moon} &= c - a \\ &= \sqrt{a^2 + b^2} - a \\ &= \sqrt{384{,}000^2\text{ km} + 1^2\text{ km}} - 384{,}000\text{ km} \\ &= 0.0000013\text{ km in one second}\end{aligned}$$

Diagram 5.9

STEP 9

APPLY THE INVERSE SQUARE LAW

The Inverse Square Law specifies that the force of gravity between two objects is inversely proportional to the square of the distance between the objects. The British polymath Robert Hooke (1635-1703) and others discovered the law, but it was Isaac Newton (1643-1727) who provided a full description in his *Philosophiæ Naturalis Principia Mathematica.*

In our example, we know (x) the distance from the center of the earth to the surface of the earth from Step 3, and we know (y) the distance from the center of the earth to the moon from Step 6. Comparing these figures indicates the distance from the center of the earth to the moon is 60 times greater than the distance from the center of the earth to the surface of the earth. Applying the Inverse Square Law (z) indicates the force of gravity, as measured from the center of the earth, diminishes by a factor of 3,600 over this distance.

When viewed from the perspective of the moon, the following statement is equally true: the surface of the earth is 60 times closer to the center of the earth than is the moon. Applying the Inverse Square Law (z) indicates the force of gravity, from this perspective, increases by a factor of 3,600 over this distance.

STEP 9

Using the Inverse Square Law to calculate the change in the force of gravity in relation to distance.

THE MATH

x = distance from center of Earth to surface of Earth
y = distance from center of Earth to Moon
z = increase / decrease in force of gravity

$$z = (y / x)^2$$
$$= (384{,}000 \text{ km} / 6{,}400 \text{ km})^2$$
$$= 60^2$$
$$= 3{,}600$$

Diagram 5.10

STEP 10

ASSESS THE RESULT

In Step 8 our calculation showed that the moon falls 0.136 centimeters in one second. In Step 9 we calculated that the force of gravity on the surface of the earth will be 3,600 times stronger than on the moon, as measured from the center of the earth. Therefore, in the same one second that the moon falls 0.136 centimeters, an object on the surface of the earth falls 488 centimeters. This is the same figure Galileo determined through direct observation of rolling balls in Step 1. Our hypothesis is confirmed.

STEP 10

Confirming that the fall of the moon and the fall of rolling balls on Galileo's inclined plane are subject to the same force of Earth's gravity.

THE MATH

Fall of Moon = 3,600 * 0.136 cm in one second
= 488 cm in one second

Diagram 5.10

If you are not amazed by this outcome, you should be. Ancients were able to measure the sizes of the earth and the moon, and by the early Enlightenment natural philosophers were able to confirm that the force of gravity causing a hammer to fall on your toe is the same as that holding the moon in its orbit.

ASK

What lessons can we draw about Reason and Religion from *The Case of Gravity*? Let's recall that the first step of the scientific method, Ask, requires a cultural openness that allows freedom of inquiry. It is therefore somewhat ironic that Galileo published his findings on the laws of motion while under house arrest on orders of the Catholic Church. His crime? Pursuing inquiries whose conclusions disputed Church doctrine.

It is easier to explain Galileo's journey from mathematician, engineer, and astronomer to prisoner of the Church by first considering his scientific predecessor, Nicholas Copernicus (1473-1543), the Polish mathematician and astronomer. Copernicus's masterwork, *On the Revolutions of the Celestial Spheres*, was published in 1543. In it, Copernicus described and argued for a sun-centered model of the universe, with the planets orbiting the sun. Conventional thinking at the time stemmed from the work of the Greco-Egyptian natural philosopher Ptolemy (c.90 CE-c.168), who placed the earth at the center of the universe.[101] In an earth-centered model, the sun and planets orbit a stationary earth. The correctness of this model was an undisputed belief in the Catholic Church. Copernicus's work thus contested entrenched doctrine, a brave act in an age when those who challenged Religion routinely were persecuted. Copernicus, however, escaped possible persecution by dying shortly before his work was published.

Nonetheless, the Church suspended *On the Revolutions*, meaning it could not be sold or read until Church-required corrections were made. The original version without corrections remained on the Church's *Index of Prohibited Books* until 1758. Martin Luther, whom we met in *The Origin of Religion*, was a contemporary of Copernicus. He reportedly labeled Copernicus a fool for challenging Holy Writ.[102] This is not surprising, as Luther saw Reason as a threat to Religion, famously stating, "Reason is the greatest whore of the devil."[103]

Returning to Galileo, in 1610 he published *The Sidereal Messenger*, a work describing his telescopic observations of the moon, Venus, and Jupiter.[104] Among his findings, he reported moons in orbit around Jupiter, which contradicted Church doctrine. The Church still subscribed to the geocentric model of the universe, the same as in Copernicus's day, 67 years earlier. Galileo's observations, to the dismay of the Church, indicated that not all celestial objects orbit the earth. To the further consternation of Church authorities, Galileo was a proponent of heliocentrism, the theory put forward by Copernicus.

In 1616 the Inquisition, a Church institution installed in the thirteenth century and charged with the eradication of heresies,[105] declared heliocentrism heretical as it explicitly contradicted Holy Scripture. Galileo was ordered to refrain from teaching or defending heliocentrism, at threat of stronger action at the hands of the Inquisition, including the all-too-real risk of torture. In 1632, however, Galileo published his seminal work, *Dialogue Concerning the Two Chief World Systems*, openly declaring the correctness of heliocentrism and refuting the arguments of geocentrism. The Church was not amused. Galileo was tried, convicted of heresy, and sentenced by the Inquisition to imprisonment; a punishment later reduced to house arrest. He remained under house arrest for the rest of his life. [106] His work on the *Two Chief World Systems* was placed on the Church's *Index of Prohibited Books*, where it remained until 1835. 360 years from its initial publication, in 1992, the Catholic Church formally cleared Galileo of any wrongdoing.

It may be tempting to view Copernicus and Galileo as exceptions in their treatment by the Inquisition and other Catholic authorities of the day. Yet they were but two of many intrepid individuals assailed and condemned for asking a disallowed question, for articulating a censurable opinion, or for challenging Holy Writ. To name other notable examples:

Abelard Pierre Abelard (1079-1142), French theologian. For championing a new model of scholarly and open discourse, Abelard was excommunicated and confined in a monastery. His books were ordered burned by Pope Innocent II (?-1143).

d'Abano Pietro d'Abano (c.1257-1316), Italian philosopher. Convicted of heresy by the Inquisition for challenging the role of spiritual influences in nature. Died in prison.

Aikenhead Thomas Aikenhead (1676-1697), Scottish student. Convicted of blasphemy for denigrating Christian doctrine. Burned at the stake.

Algerio Pompino Algerio (1531-1556), Italian law student. Condemned by the Inquisition for arguing the right to attend any church. Executed by boiling in oil.

d'Ascoli Cecco d'Ascoli (1257-1327), Italian physician and mathematician. Convicted of impiety by the Inquisition. The first university scholar to be burned at the stake.

Bruno Giordano Bruno (1548-1600), Italian Dominican friar and philosopher. Convicted of heresy by the Inquisition, due in part to his scientific beliefs. He believed the universe was infinite and stars had their own planets. Burned at the stake.

Campanella Tommaso Campanella (1568-1639), Dominican friar and philosopher. Denounced to the Inquisition for arguing against the teachings of Aristotle. Tortured and imprisoned.

Cardano Gerolamol Cardano (1501-1576), Italian mathematician and physician. Arrested by the Inquisition for reasons unknown and imprisoned.

Cranmer Thomas Cranmer (1489-1556), English Archbishop. Convicted of heresy for his participation in the English Reformation. Burned at the stake.

Dolet Étienne Dolet (1509-1546), French translator and printer. Accused of atheism by the Catholic theological faculty of the Sorbonne. Executed.

Erasmus Erasmus of Rotterdam (1466-1536), Dutch humanist and theologian. For challenging scriptural interpretations and accepted doctrine, his works were placed on the *Index of Prohibited Books* by Pope Paul IV.

Hamont Matthew Hamont (?-1579), English ploughwright. Convicted of heresy by an ecclesiastical court for proclaiming scripture a fable. Burned at the stake after having his ears cut off.[107]

Hus Jan Hus (c.1369-1415), Czech priest and philosopher. Argued for reforms within the Catholic Church. Declared a heretic, excommunicated, and burned at the stake.

Hypathia Hypathia of Alexandria (c.350-415), Greek mathematician and astronomer. Accused of pagan and non-Christian beliefs. Sided with a Roman governor against the Bishop of Alexandria. Murdered by a Christian mob.

Lyszczynski Kazimierz Lyszczynski (1634-1689), Polish nobleman. Accused of atheism by the Bishop of Kiev. Executed.

Servetus Michael Servetus (c.1509-1553), Spanish polymath. nobleman. Disputed infant baptism and the doctrine of the Trinity. Charged with heresy by John Calvin, convicted, and burned at the stake.

Telesio Bernardino Telesio (1509-1588), Italian philosopher and scientist. For advocating an understanding of the world based on observation rather than abstract reason, his works were placed on the *Index of Prohibited Books*.

Toland John Toland (1670-1722), Irish rationalist and author. Condemned by Catholic members of the Parliament of Ireland for arguing that faith can be understood by reason. His books ordered burned.

Vanini Lucilo Vanini (1585-1619), Italian philosopher and physician. Convicted of heresy and atheism by the Parlement of Toulouse, which operated the Toulouse Inquisition. Strangled at the stake.

de Viau Théophile de Viau (1590-1626), French dramatist and poet. Accused of atheism by the Jesuits. Ordered burned at the stake, but escaped.

Woolston Thomas Woolston (1668-1733), English theologian. Argued against the literal interpretation of scripture. Convicted of blasphemy and imprisoned until his death.

Wycliffe John Wycliffe (c.1325-1384), English philosopher and theologian. Reformer who challenged the privileged status of clergy. Declared a heretic by the Council of Constance in 1415. Posthumously excommunicated.

Thomas Aikenhead is perhaps the most notable entry on the list. He is notable for not being notable. He possessed no scholarly credentials, occupied no position of prominence, and authored no magnum opus. He was a student. He was a commonplace. His crime was to express disbelief - disbelief in the fables of the Old and New Testaments. He referred to Holy Scripture as arrant nonsense. On the afternoon of 8 January 1697, he paid for his opinion with his life.[108]

The Inquisition, as the Catholic Church's principal arm of enforcement, adopted different forms in different countries, but in all cases operated under the authority of the Church or Christian rulers. It increased in size in response to perceived threats, most notably during the Protestant Reformation. Its existence and its exploits instilled anxiety, terror, and panic among its targets. Meticulous in its mission, the Inquisition conducted trials and required proof of guilt. The proof, however, was obtained by confession, often coerced through torture. The legendary confessions extracted from the Knights Templar exemplify the zeal of the inquisitors.

The Templars, a Christian military order endorsed by the Catholic Church in 1129, rose to renown with military victories in the Crusades. The organization grew in wealth and power over the next two centuries, exercising all but extraterritorial authority, with interests in banking, shipping, manufacturing, construction, real estate, and international trade. In 1307 Pope Clement V, anxious to reduce the influence of the Templars, conspired with King Philip IV of France (1268-1314) to plot their downfall. Philip, heavily in debt to the Templars, became a willing conspirator. At dawn on a Friday morning in October of the same year, state authorities arrested all the Templars in the country on spurious charges of heresy, relegating their prisoners to the Inquisition.

To extract confessions, per Tuchman, the Templars were "racked, thumb-screwed, starved, hung with weights until joints were dislocated, had teeth and fingernails pulled one by one, bones broken by the wedge, feet held over flames, always with pauses in between and the 'question' put again each day until confession was wrung or the victim died."[109] Thus were confessions obtained and guilt confirmed.

There are no reliable statistics on the number of individuals who were investigated and condemned by the Inquisition. Estimates run into the tens of thousands. The Inquisition set its agents to work against those who posed the wrong question, professed the wrong belief, or presented a threat - in the eyes of the Church - to Catholic authority. The institution assiduously pursued its remit over six hundred years. In 1904 it was absorbed into other offices of the Church, where it continues to reside today as the *Congregation for the Doctrine of the Faith*.

In preceding paragraphs, we referenced the *Index of Prohibited Books* as an instrument to suppress unwelcome inquiries and muzzle undesirable ideas. The *Index*, which was created in 1559 by order of Pope Paul IV (1476-1559) and periodically updated over the next four centuries, contained the works of writers and thinkers from a variety of disciplines: mathematicians, astronomers, physicists, philosophers, novelists, poets, polemicists, and theologians. Well-known names whose works were placed on the *Index* include Blaise Pascal, David Hume, Francis Bacon, Immanuel Kant, Jean-Jacques Rousseau, Jean-Paul Sartre, Johannes Kepler, John Locke, John Milton, Galileo Galilei, Nicolas Copernicus, René Descartes, Victor Hugo, and Voltaire. Inclusion limited the printing, distribution, and teaching of the restricted works. Imagine attending a religious educational institution that suppressed an untold number of the most eminent ideas of the age. The *Index* was abolished in 1966 by Pope Paul VI (1897-1978), not as a relic of an unenlightened age, but because it became too arduous to maintain in an era of widespread literature.

Copernicus and Galileo typify an age where Religion prohibited the asking of questions that challenged existing beliefs. Yet the Catholic Church was far from alone in this practice. Over the centuries, Christian and non-Christian denominations of all stripes deliberately created cultures of intolerance that discouraged inquiry. Each employed threat, excommunication, torture, and execution to reinforce the message. Each further engendered an environment of self-censorship, where personal guilt - the remorse bound up with challenging an ingrained conviction - deterred inquiry. To what end? To preserve the dogmatic truths of a preferred faith, to impose the truths on all concerned, and to ensure the truths remain both unquestioned and unquestionable.

And what of today? Has Religion abandoned its chronicled resistance to inquiry? Has it relegated its past behavior to the dustbin of history? Not at all. Its aversion to inquiry continues. In too many parts of the world, Religion still opposes the asking of undesirable questions. It correctly interprets inquiry as a threat. Inquiry challenges dogmatic authority. Inquiry exposes doctrinal falsehoods. Inquiry uncovers scriptural inconsistencies. Inquiry reveals fallacious reasoning. To all of which Religion continues to respond with time-tested tropes. It elevates faith beyond the realm of inquiry. It cultivates a culture inhospitable to investigation. And when pressed, it shames, ostracizes, or otherwise punishes the inquirer. Witness laws against heresy, blasphemy, and apostasy throughout the Middle East, North Africa, Southeast Asia and elsewhere.

The rise of conservative Islām, in stark contrast with Islām's Golden Age, provides but one specific example. The Nobel Prize-winning Pakistani physicist, Abdus Salam (1927-1996), berated conservative Islām's hostility to inquiry in particular and science in general, by declaring, "Of all civilizations on this planet, science is weakest in the lands of Islām. The dangers of this weakness cannot be over-emphasized, since the honorable survival of a society depends directly on its science and technology."[110] His viewpoints and beliefs earned him a standard-issue recompense: excommunication, this time from the Ahmadi, the Muslim denomination to which he subscribed. Pervez Hoodbhoy (1950-), another Pakistani physicist and a former professor at Quaid-i-Azam University in Pakistan, offered another insight on the prejudicial attitude to inquiry. He lamented the university had several mosques but no bookshop.[111]

There are instances in Reason, of course, where a well-established theory hinders inquiry, simply because broad acceptance of the theory seems to make further investigation unnecessary. Newton's theory of gravitation, as previously noted, remained unchallenged for over two centuries because it was thought to be complete. In other cases, influential scholars sometimes employ academic invective to squelch uncomfortable questions, often to protect their own theories and reputations. The translation of Maya hieroglyphs in the twentieth century illustrates.

The Maya civilization of Mesoamerica employed a hieroglyphic script from the third century BCE until the Spanish conquest in the sixteenth and seventeenth centuries CE, at which time the faculty to read the script was lost. Nearly three centuries later, the English archaeologist J. Eric Thompson (1898-1975) studied a handful of surviving Maya books and stone inscriptions, seeking to decipher the script. He determined that Maya hieroglyphs were logograms or ideograms - symbols conveying meaning but not sound. In English, for example, the symbol 2 is an ideogram, as it represents meaning (denoting the quantity two) but commands no phonetic significance. A later, competing theory argued that Maya hieroglyphs denoted a phonetic, alphabetic script. Thompson, the eminent scholar in his field for five decades, publicly and harshly criticized any scholar who supported the competing theory. Thompson's authority and strength of personality effectively delayed the full decipherment of the Maya language until 1973, at which time a preponderance of evidence forced Thompson to concede his position.[112]

A postscript to this story underscores the contribution of Religion to the translation of the Maya script. At the time of the Spanish conquest, Diego de Landa (1524-1579), a Spanish priest and later bishop of the Catholic Church, declared Maya books to be pagan and ordered them burned. Today, only four Maya books survive.[113]

In Reason, when a well-accepted theory or an eminent scholar inhibits inquiry, the setback frequently is short. And, more important, it never is punctuated by torture. Reason, in contrast to Religion, never possessed an Inquisition to persecute heretics, never conducted an auto-da-fé to burn proponents of geocentrism, and never impaled anyone for suggesting planets orbit in perfect circles. The same is true in today's world. Reason punishes no one for asking unwelcome questions. It murders no one in Bangladesh[114] for the "sin" of atheism, disappears no one in Pakistan[115] on the accusation of apostasy, imprisons no one in Saudi Arabia[116] for publishing an irreligious blog, fines no one in Singapore[117] for insulting theism, and discriminates against no one in the USA[118] for adopting a gender identity contrary to "God's will." Religion can assert no similar claims.

PREDICT

We now move to the second step of the scientific method, Predict. What lessons does *The Case of Gravity* offer for Predict?

Testable and falsifiable predictions, as noted in *The Origin of Reason*, form the most important element of the scientific method. Every step articulated at the start of this chapter offers such predictions, each confirmed or disproved at the time of the prediction or at a later date employing improved technology. In Step 1 (determining the force of gravity on Earth), for instance, Aristotle formulated a testable prediction that falling objects travel the same distance in successive time intervals, a prediction disproved by Galileo. In Step 5 (determining the distance to the moon), modern laser ranging confirmed the distance predicted by Eratosthenes. You likewise have the ability to independently test the hypotheses and outcomes, by constructing your own inclined plane or by replicating the technique to measure the distance to the moon. You are not forced to accept the word of Galileo, Eratosthenes, or the author of this book. Such transparency is an integral element of Reason. Galileo and Eratosthenes published their works. Their methods are publicly available. They do not rely on enigmatic proclamations, cryptic pronouncements, private revelations, or anagogic interpretations.

The Case of Gravity underscores a second aspect of Predict: the necessity for detail. Predictions must incorporate granular, fastidious, and sharply defined detail. Detail conveys understanding. Detail imparts confidence. Detail is the foundation of testability. Without detail, subsequent experiment cannot confirm or disprove anticipated outcomes. Again, all ten steps outlined at the start of this chapter contain sufficient detail for testable predictions. Indeed, a motivation for recounting each step was to demonstrate the importance of detail.

Hypotheses that lack detail, sometimes referred to as *black box* hypotheses, exist in two forms. The legitimate form presents the black box as an abstraction, expressing a relationship between inputs into and outputs from the box. While current research may focus on the inputs and outputs, future investigation attempts to explain the internal workings of the box. The illegitimate form, in contrast, simply accepts the black box, ascribing its interior workings to mysteries beyond human comprehension. Hypotheses that omit detail and unquestioningly accept the existence of mystical black boxes are invalid.

How does Religion develop its predictions? To answer this question, we start with William Miller. Miller (1782-1849) was a farmer and lay preacher. Around 1833 he initiated a national movement of 50,000 to 500,000 followers (estimates vary) in the United States named, eponymously and unsurprisingly, Millerism, an offshoot of Protestant Christianity. Miller predicted the Second Advent of Jesus Christ would occur on or before 1843. He based his prediction on a prophecy in the Bible which refers to the cleansing of the sanctuary in "two thousand and three hundred days."[119] Miller deciphered the word *days* to denote *years*, and believed the prophecy originated in 457 BCE. Thus, 457 BCE + 2,300 years = 1843 CE.[120] When the predicted date duly arrived, Jesus did not. Miller then revised his prediction to April 1844. The result was the same. Miller concocted a third prediction, this time that Christ would return in October 1844. You know what transpired. Incidentally, despite repeated failures, his followers did not desert their faith. In 1863 they went on to found the Seventh Day Adventists, another denomination of Christianity.

Miller's attempt at prophecy was not unique. History is replete with predictions of the second coming of Christ, the destruction of the world, and other religiously significant events. Christianity, throughout the two millennia of its history, has offered any number of such predictions, both divine and secular. We already discussed one of these, the belief that the sun revolved around the earth. Yet Miller's formulation yields an archetype for our narrative. How so? He articulated a testable religious prediction, one that could be unambiguously confirmed or disproved. A sampling of such prophecies - each unequivocal, each articulating a specific result - yields a pattern that allows us to answer a key question. What is Religion's record in formulating successful predictions? Let's find out by examining three classes of religious predictions: cryptic, explicit, and derived.

Cryptic predictions appear in scripture but are vaguely expressed. They require interpretation. Think Millerism.

Explicit predictions also appear in scripture. But unlike their cryptic cousins, they are articulated in plain language: clear-cut, unambiguous, and often literally embraced.

Derived predictions rely on non-scriptural sources to supplement Holy Writ. Clerical pronouncements and theological treatises typify this class of predictions. We will deconstruct the three classes in reverse order.

DERIVED PREDICTIONS

To illustrate derived predictions, we turn to the Italian theologian and philosopher, Thomas Aquinas (1225-1274). Aquinas's unfinished masterwork, *Summa Theologiæ,* written between 1265 and 1274, explains the salient teachings of Christian theology. It also attempted to reconcile aspects of Aristotle's natural philosophy with both Christian doctrine and the reality of God's perfection. It is presented as a series of over five hundred articles, each following a set structure: Aquinas states a conjecture (one he deems false), enumerates objections to an unstated conclusion, presents his own counterargument, then replies to the original objections.[121] It is a beautifully-written, well-reasoned, scholastic work. Aquinas clearly was a clever fellow.

It is the success of Aquinas's labors that raises the issue in this narrative. That success concerns Aristotle. Aristotle was the most influential scholar of his age, his teachings influencing both Islām and Christianity. In the West, his philosophical and scientific writings shaped European thinking for over 1,000 years, well into the early Renaissance. The problem is that significant elements of Aristotle's science were incorrect. He often reached conclusions via reason (with a small *r*) rather than from observation. Indeed, as noted earlier, he is the father[122] of deductive argument by syllogism.[123] He believed, for example, in geocentrism, a topic we already visited.[124] Among his writings, he also suggested that stars obtain their light and warmth not from internal sources, but from friction as they move through heavenly air;[125] that stars move not on their own, but are attached to celestial circles that revolve around a common center;[126] and that the firmament contains a divine substance (an aether) distinct from earthly elements.[127]

With Aquinas's synthesis of Aristotelian and Christian thought, Christianity came to incorporate the incorrect beliefs of Aristotle, albeit with a theological twist. God is perfect, a circle is the most perfect form, ergo the orbits of the planets are circular. It probably was not realized at the time, but this is one of a host of testable predictions that Christianly adopted. Through this process, Christianity affirmed the verity of a universe awash with invisible aether,[128] of a celestial firmament eternal and unchanging,[129] and of planets and stars orbiting in circles[130] at uniform speeds.[131]

Aquinas, though notably influential, was but one of many theologians who inspired Christianity to adopt a miscellany of testable predictions. Many were inscribed as divine doctrine, affirmed as everlasting truth, and promulgated to the faithful. To challenge their veracity constituted heresy. Yet what epitaph can we ascribe to the predictions? Two words - widespread refutation. Stars orbit in ellipses, not celestial circles; planets move at variable speeds, not fixed velocities; the celestial heavens are dynamic, not static; and there is no aether. This pattern, from ancient times to present day, is unvarying: prediction after prediction unceremoniously disproved. The reason is clear. Religious predictions embody no divine wisdom from revelation, scripture, or prayer. They simply arise from wishful thinking and the shopworn lore of their era. Nothing more. Perhaps Aquinas realized this. In his *Summa,* he articulates his Five Ways to "prove" the existence of God.[132] Not one offers a testable hypothesis.

EXPLICIT PREDICTIONS

Explicit prophecies, as a reminder, express unambiguous scriptural claims in plain language. Here, consider a passage from the Bible, which asserts that its deity stopped the sun, providing Israelites with extended daylight to avenge themselves upon their enemies:

"Sun, stand thou still upon Gibeon ... So the sun stood still in the midst of heaven, and hasted not to go down about a whole day."[133] Here we have an explicit, unambiguous assertion - the sun stood still. How might Religion substantiate this claim? For openers, Religion could acknowledge its deity misspoke. Stopping the sun's rotation or its orbit around the Milky Way galaxy would not alter the quantity of light falling on the earth. It is odd that a supposedly omniscient deity would base his deed on the disproved theory of geocentrism. So perhaps the deity meant he had stopped the rotation of the earth to preserve sunlight on one side (the Israelite side) of the planet. What are the consequences of such a rotational stoppage? Let's consider four of the many repercussions.

Gravity, combined with the centrifugal force of the planet's rotation, causes the oceans to bulge at the equator. Would stopping the earth's spin send equatorial waters to higher latitudes, creating global, tsunami-style floods? What about the atmosphere, which is traveling over 1,600 kilometers per hour at the equator? When the solid earth stops spinning, would the atmosphere continue moving, scouring the surface of the planet with supersonic winds at 10-15 times the potency of a hurricane? What of the earth's magnetic field, which is thought to be linked to a dynamo effect caused by the planet's rotation? Would the field decay or collapse, exposing all life forms to deadly solar wind and cosmic rays? Or did the earth completely stand still, as the passage from Joshua implies, denoting it also stopped orbiting around the sun? If the earth suddenly ceased traveling at 100,000 kilometers per hour around the sun, would it immediately begin to plunge into the sun, its atmosphere vaporizing as it approached the corona (whose temperature is above one million degrees Celsius) of our home star? The point should be clear. Physical processes are inextricably linked. Altering one creates a chain of consequences. The authors of the Joshua parable, informed only from the crude knowledge of their era, possessed no awareness of such interdependencies.

How might Religion substantiate its claim of a deity stopping the sun in the sky? By providing detail. Testable, falsifiable detail. Yet Religion supplies none, offering only a perfunctory *Goddidit* - a vacuous catchphrase ascribing as explanation of a phenomenon nothing more than the swish of a magic wand. Religion then dissembles. When confronted with the fact that a religious claim contradicts all known physical reality, Religion seeks an escape route. In our example, by declaring that the Biblical passage narrates no physical event, but employs a literary device to convey a broader, abstract message. The passage thus becomes an allegory, a hyperbole, a *Deus ex Machina*. Note the stratagem. After centuries averring, with furrowed brow, the literal truth of the passage, Religion suddenly and unashamedly advocates a poetic interpretation. How convenient.

Similar stopping-the-sun (or stopping-the-earth) contentions pervade the Bible and the writings of church fathers: a sea magically parted to allow crossing on dry land;[134] a river divided for an analogous purpose;[135] a horde of devil-infested swine committing suicide by drowning;[136] a man surviving the depths of the ocean in the stomach of a fish;[137] an ax with the ability to swim;[138] a plague of frogs supernaturally enjoined then summarily dispatched;[139] and the confirmation of a fiery purgatory awaiting the unrepentant sinner.[140] For hundreds of years, the lion's share of its history, Christianity avowed all as literal, unassailable truths, persecuting anyone who dared challenge their veracity. Yet now, its predictions both rebutted and discredited, Religion adopts a new tack. It handily and abjectly adopts a figurative truth. And, markedly, it adroitly transforms testable predictions into non-testable pretenses.

CRYPTIC PREDICTIONS

Cryptic prophecies, our third class of predictions, rely from the outset on scriptural interpretation. This time, note a passage from the Qu'rān, which many an Islāmic scholar cites as a verified religious prophecy: "We have certainly made the Qu'rān easy for remembrance"[141] The verse, as interpreted by theologians, asserts the Qu'rān was endowed by Allāh to abet memorization, and uniquely so. The fact that many have memorized the tome is deemed proof of the assertion. Thus is prophecy confirmed.

Yet the practitioner of Reason, dumbfounded by the claim, quickly discerns innumerable sophistries entrenched in both the assertion and its confirmation. To name but a few. One, the claim exploits the "same fact twice" ploy described in *The Origin of Reason*. Two, no empirical data are cited. What experiments support the claim that no other text is as easily memorized? Three, the claim cherry-picks evidence. If a similarly structured text of comparable length is memorized, say, a science fiction book, is it also a sign of divine intent? Four, the claim is not falsifiable. Does failure to memorize the text refute its divinity? Five, the claim tenders no novel predictions. Memorization in five minutes of a previously unseen text would be novel, and it could be easily arranged by an omnipotent deity. Has it happened?

Six, the claim suffers from logical fallacies, a topic we cover in *The Case of Earth's Age*. One such fallacy is that of *questionable authority*, where an assertion from an authority is likely to be biased. In this instance, the claim that theistic intent permits ease of memorization arises from a partisan, religiously motivated source: Islāmic theologians and imāms. And seven, there is a confusion of cause and effect. What is the role of the broader community in encouraging the literal recitation of one specific book, to the exclusion of all others? How does exposure to extracts from the text, publicly and privately voiced, day in, day out, year after year, affect recollection? How do university courses, specifically designed to teach the text and promote its recall, influence mastery? How do state-sponsored competitions, which offer lucrative prizes for reciting the Qu'rān, affect one's motivation to learn the scripture? We'll conclude the list here; you grasp the point. The vagueness of enigmatic claims, their imaginative interpretations, their lack of detail, their absence of objective data, their omission of novel predictions, their non-falsifiable constructs - all serve to invalidate their prophecies and their putative confirmations.

Cryptic, explicit, derived, and other classes of predictions pervade not only Christianity and Islām, but the doctrines of all faiths. Yet all such prophecies share a common hallmark, one exemplified from the preceding examples. It is, once again, this: Religion's record of formulating successful predictions is appalling. Faith-based pronouncements - each founded on revelation, scripture, or prayer - afford no expository or predictive power. None.

Reason, too, suffers from a lengthy history of testable predictions that have failed. The distinction, or course, is that Reason does not claim eternal and immutable truth. The scientific method produces accepted theories based on an extensive range of observational evidence and successfully-tested predictions. But Reason is open to refining or rejecting a theory when new, contradictory evidence arises. Reason recognizes that knowledge and understanding progress gradually, in the company of occasional reverses. It likewise comprehends that every claim revealed to be false narrows the venues where truth might be found. Religion's compass, by contrast, is unidirectional. By claiming immutable truth, it is difficult for Religion to alter course. It is arduous (and problematic) for Religion to say, "Oh, that eternal truth I told you about yesterday, well, uh, forget it, today it's different."

As Reason advances, it becomes increasingly capable of refuting Religion's testable affirmations. How does Religion respond to this predicament? How does Religion defend against the continual disproof of its claims? What lessons has it absorbed from the untold refutations of its assertions, many similar to those cited in this chapter? At this point in our narrative, four ruses stand out.

First, Religion shuns detail. It eschews reductionism - the idea that complex phenomena can be understood by the granular and rigorous examination of their component parts. It thus renounces precisely those ingredients demanded for clarity of thought, for meticulous experimentation, and for sharp differentiation of results. Its predictions instead rely on mystical black boxes, each disclosing only vague and illusory interpretations of both earthly and otherworldly events. This mindset, one disinclined to energetic inquiry, concludes that no measure of effort can (or should) decipher the mechanisms within the box. To what purpose? All too often, to instill an incurious mentality, to terminate further inquiry, to avoid possibly unwelcome detail, and to advance self-serving interpretations. Absence of detail and mystical black boxes serve to deprive Reason a foothold, lowering the risk a religious claim will be refuted.

Second, as noted from the stopping-the-sun assertion, Religion refashions refuted claims to restore alignment to prescribed doctrine. Theologians lay hold of enshrined truths, now refuted but long-proclaimed as the literal word of god, and craftily reinterpret them as figures of speech or edifying parables. In this manner, Religion gives itself an "out," allowing it to cling to its obdurate, dogmatic beliefs. To the devout, Religion rehabilitates its tales subtly and without ignominy. To the objective observer, such contrivances discredit Religion and further evince its inability to predict or explain phenomena. And to the countless souls persecuted in ages past for challenging the literal canon, Religion expresses little remorse.

Third, Religion has come to avoid the creation of any and all testable predictions. This is perhaps Religion's greatest learning. To preclude refutation, Religion frequently and deliberately evolves its beliefs to ensure they are invulnerable to observational experiment. [142] It is not merely a matter of awaiting a future date, at which time new techniques and technologies might permit the testing of a prediction. Instead, Religion frames its pronouncements so they can never be invalidated, thus neutralizing the power of inductive reasoning. It has to, as Religion possesses no comparable capability to the scientific method. To quote Carl Sagan, "There isn't a religion on the planet that doesn't long for a comparable ability - precise, and repeatedly demonstrated before committed skeptics - to foretell future events."[143]

A beautiful instance of the non-testability of religious proclamations was articulated by the British philosopher Bertrand Russell (1872-1970). It is called Russell's teapot.[144] Russell placed his teapot between Earth and Mars. I place it behind your head. I claim there is a teapot floating right behind and just above your head. It has always been there and always will be, throughout your entire life. It is a beautifully painted, pale blue, ceramic teapot. It happens the teapot is endowed with a unique property: You cannot sense it in any way. If you turn your head, the teapot shifts its position, remaining just outside your field of vision. If you look in the mirror, the vessel reflects no light. If you reach behind your head and attempt to strike the teapot with a hammer, it vanishes, only to reappear when the danger has passed. No empirical test discloses its existence. No one else can sense its presence. Yet it floats at this very moment, just behind your head.

Though not as eloquently presented as Russell's, my teapot claim is equally untestable. Deliberately so. This was Russell's contention, and this is the practice of faith-based thinking. Religion studiously constructs its doctrines to thwart testability. John Bagnell Bury (1861-1927), the Irish historian and philosopher, concluded the analogy thus: "Some people speak as if we were not justified in rejecting a theological doctrine unless we can prove it false. But the burden of proof does not lie upon the rejecter."[145] This is a critical point, reaffirming the quote at the beginning of this chapter. When an extraordinary claim is presented, the onus is not on the skeptic to disprove the claim. The onus rests with the claimant to prove its validity. The onus is on Religion to prove the validity of its revelations, its prophecies, and its deities. The teapot, as emblematic of a non-testable claim, is as valid as any non-testable religious belief. And just as worthy of dismissal.

Of the four lessons Religion has embraced to assuage the disproof of its assertions, we now have recounted three: It eschews detail, recasts its narratives, and evades testable predictions. Here is the fourth. Religion responds to the concept of testable predictions by attempting validation through faith. Faith, of course, derives from the five sacreds, a topic we covered in chapter two. Behind faith stands divine authority. From this authority, the follower is expected to blindly accept any and all pronouncements - each attended by a comprehensive absence of evidence - with an enthusiastic *Amen* or *Inshā'a Allāh*. But consider the ramifications of belief in an authoritative, tri-omni deity. He can recast the laws of the universe on a whim, his revisions interpreted by the devout as miracles or punishments. There are no rules, no laws, no structures comprehensible by mere *H. sapiens*. Anything can happen at any time, rendered under the guise of divine inconstancy. Conveniently for Religion, testable predictions serve no purpose in such an arbitrary universe.

We arrive at the end of *The Case of Gravity*. To recap, our exploration of Ask illustrated that Reason embraces a culture of inquiry. Reason encourages, even demands, the asking of questions without regard to existing beliefs or deeply rooted convictions. Reason ranks inquiry as the critical first step in the search for new knowledge, as the catalyst that propels a more comprehensive understanding of the world. It is the failure to ask questions that threatens Reason.

Religion, by contrast, often discourages and even punishes questioning. It appeals to authority to suppress distasteful ideas and terminate unwanted inquiry. It recognizes (and fears) that inquiry will open gaps and expose inconsistencies in its theology, endangering its entire chimerical edifice. At its worst, Religion applies ostracism and threats of violence to muzzle inquiry, preferring unquestioning ignorance as the cornerstone of its notional dogma. The Italian philosopher, Cesare Cremonini (1550-1631), further illustrates this aspect of Ask. When offered the opportunity to look through Galileo's telescope, he refused.[146] Out of fear of the Inquisition, he chose not to see evidence that would contradict the teachings of the Catholic Church.

Our exploration of Predict demonstrated that Reason bases its *how to think* process on testable predictions, which constitute - to repeat yet again - a nonpareil ability to discern fact from myth. Religion, possessing no equivalent capability, finds its predictions routinely disproved. It counters by reframing its ideology to impede testability, and by championing fideism - the concept that faith exists outside the realm of testability, standing apart and superior to Reason. It propagates a generic definition of faith as the belief in something, anything, in the total absence of supporting evidence. It then proudly waves the banner of faith as incontrovertible truth. In this context, faith is an invented construct, formulated to perpetuate allegiance to mythical entities with no pretext of rationality, serving solely as a refuge from the uncomfortable truths of Reason. Perhaps the dictionary definition of faith should be *a belief intentionally designed to thwart disproof.*

Let's spend another moment on the word *faith*. Religion sometimes accuses Reason of invoking its own *faith*, as when a scientist says, "I have faith the experiment will deliver the expected results." But this is erroneous. First, the word *faith* in this context denotes *confidence* rather than *blind belief*. As such, its use reflects sloppiness of language. Second, even when confidence reflects faith in an outcome, it is an informed faith relying at the very least on a modicum of evidence rather than cabalistic explanations. It is not the ingrained, unquestioning faith of Religion. Third, and most notably, if the experiment fails to deliver the expected results, the practitioner of Reason will accept the evidence and alter her *faith* accordingly, a behavior rarely demonstrated by Religion.

The Case of Gravity highlights a further crucial distinction between Religion and Reason. In Religion, a deity presents an immutable truth, which is meant to be eternal and accepted generation after generation. In Reason, truth is a journey, each discovery forming a springboard for the next discovery. The accurate astronomical observations of the Danish astronomer Tycho Brahe (1546-1601), for example, enabled the German astronomer Johannes Kepler (1571-1630) to formulate the laws of planetary motion. Kepler's work, in turn, influenced Newton's laws of motion and his formulation of universal gravitation. Newton's experiments on light led to the successful unification of magnetism and electricity by the Scottish physicist James Clerk Maxwell (1831-1879). Maxwell's work then influenced Einstein's theory of special relativity.

And so the chain continues - thousands upon thousands of individuals, each contributing to Reason's body of knowledge, each discovery shining a light on the next mystery, each enigma providing a focus for the next iteration of discoveries. Some breakthroughs emanate from a select few, an intellectual elite aberrantly endowed with the ability to espy and decipher the complexities of the world.

Yet the majority of advances derive from practical knowledge, by expertise secured through dint of hard work, by the curious mind driven by utilitarian function. History is replete with examples. To cite a few of the many: by blacksmiths who understood the properties of metals; by road builders with insights on geology; by land surveyors with observations on geography; by midwives with an acumen of childbirth; by agricultural breeders with knowledge of trait inheritance; by ship captains refining navigational instruments; and by laboratory technicians classifying bacteria from their morphology. Science, as such, is every bit a bottom-up as a top-down endeavor.[147] Often more so.

We close this chapter with a final observation on Copernicus. By placing the sun rather than the earth at the locus of the Solar System, Copernicus launched the first great displacement of humanity. Humanity was no longer at the physical epicenter of the universe, no longer at the physical nexus of a deity's creation. A second, equally remarkable displacement would come 316 years later, a topic we cover in a future chapter.

Diagram 5.12 presents the conclusions from *The Case of Gravity*.

MISCELLANY

In the inclined plane experiment, the measurement of a ball falling 488 centimeters in one second appears in no Galilean text. There are two reasons for this. First, Galileo conducted his experiments prior to the invention of the metric system, recording his measurements in pounds (weight) and cubits (length, roughly the distance from the elbow to the tip of the middle finger). Second, Galileo employed an inclined plane specifically to slow the descent of rolling balls, in order to measure accelerated motion. Adjusting for the angle of the ramp, Galileo's findings equate to a vertical fall of 488 centimeters in the first second. As an aside, Galileo reports his ramp was 12 cubits in length, about 5.5 meters. The top end of his ramp was elevated two cubits higher than the lower end. Some accounts suggest Galileo notched marks on the ramp to measure distance rather than using bells.

I use the term *Inquisition* loosely. The Catholic Church operated the institution in multiple forms, across diverse geographies, and under various names. To cite several: the Papal Inquisition, the Medieval Inquisition, the Roman Inquisition, the Portuguese Inquisition, the Supreme Sacred Congregation of the Roman and Universal Inquisition, the Supreme Sacred Congregation of the Holy Office, the Congregation for the Doctrine of the Faith, or simply the Holy Office. Pope Gregory IX (c.1145-1241) established the first Inquisition in 1232. The more widely known Tribunal of the Holy Office of the Inquisition (aka the Spanish Inquisition) was established in 1478. The Church extended its reach to the Americas as well, initiating the Mexican Inquisition in the late sixteenth century.

Aquinas clearly admired Aristotle, mentioning him well over 1500 times in his *Summa*, either by name or as "the Philosopher." This reverence not only implanted Aristotelian thought as a legitimate source of Christian truth, but empowered other theologians to patronize the same and similar founts. Here is a further sampling of testable predictions from Aristotle and other ancient sources. The universe is spherical.[148] Stars are indestructible.[149] Planets orbit perpetually.[150] The speed of light is instantaneous.[151] A lighter body of water exists above the celestial firmament.[152] Each such prediction, assimilated into the Christian faith and tenaciously defended for centuries as divine truth, was ultimately disproved.

Aquinas' Five Ways to prove the existence of God rest on arguments concerning first motion, efficient cause, necessity, gradation, and design.

A summary of the different approaches of Religion and Reason in the first two steps of the scientific method, Ask and Predict.

Diagram 5.12

Principles

	Religion	Reason
① ASK	Compels adherence to dogma; suppresses, even punishes, inquiry that challenges existing beliefs and convictions.	Encourages a culture of openness, advocates inquiry that challenges established knowledge.
② PREDICT	Deliberately structures beliefs to avoid testable predictions, posits faith as superior to evidence.	Demands hypotheses with testable predictions that can be confirmed or disproved by observation and experiment.
③ OBSERVE	To be discussed in Chapter VI.	To be discussed in Chapter VI.
④ VERIFY	To be discussed in Chapter VI.	To be discussed in Chapter VI.
⑤ ACCEPT	To be discussed in Chapter VII.	To be discussed in Chapter VII.

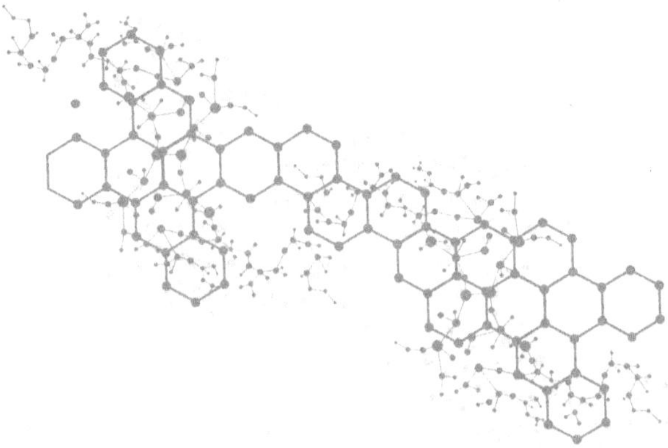

[VI]

The Case Of Earth's Age

No evidence is powerful enough to force acceptance of a conclusion that is emotionally distasteful.[153]

THEODOSIUS DOBZHANSKY (1900–1975) RUSSIAN-AMERICAN BIOLOGIST

The Protestant Archbishop James Ussher (1581-1656), after extensive Biblical study, declared that creation began in 4004 BCE. More precisely, construction of planet Earth began on Monday, the twenty-third of October.[154]

Ussher's research provided a long-sought answer to one of humanity's elementary questions: How old is the earth? This is the riddle we examine in *The Case of Earth's Age*. Both Reason and Religion have produced a variety of dates for the origin of the planet. We will probe the various answers and, of equal import, how each domain's *how to think* process produced the answers. For Reason, *The Case of Earth's Age* also will allow us to take a closer look at the third and fourth steps of the scientific method, Observe and Verify.

Reason offers several techniques for dating the age of objects. One of the more familiar methods resides in the field of dendrology, the study of woody plants. A sub-discipline, dendrochronology, refers to the use of tree rings for dating. It is well-established that rings form in many types of wood and can be dated to specific calendar years. A broad range of samples improves the reliability of dating.

While some trees might live for 3,000 to 5,000 years, it is more common to date older objects by examining shorter-lived trees in an overlapping manner. In Diagram 6.1, the newest wood (***a***) might come from a recently cut tree. The intermediate wood (***b***) might be a beam from an old house. The oldest wood (***c***) might have been obtained from another, earlier structure. By examining overlapping ring patterns, dendrochronology provides reliable dating back to 11,000-12,000 years.[155] If a more exhaustive set of unbroken, overlapping ring patterns from older timbers is someday discovered, dendrochronology might be extended further back in time.

Another well-known technique for determining the age of objects is radiocarbon dating. Carbon is one of the 92 naturally occurring elements in the Periodic Table, a list of all known elements. Carbon occupies position number 6 in the table. The 6 refers to the structure of the carbon atom.

Atoms contain positively charged particles named protons, and an equal number of negatively charged particles called electrons. The number of protons in an atom denotes the *atomic number* of the atom, which in turn denotes the type of element. An atom with 6 protons is atomic number 6, which is carbon. Subtract one proton from the atom, and you have boron, atomic number 5. Add a proton to carbon, and you have nitrogen, atomic number 7. Most atoms also contain uncharged particles named neutrons. The protons and neutrons reside in the nucleus of the atom, with the electrons orbiting the nucleus in shells or layers as depicted in Diagram 6.2. This view is simplistic but serves our purpose.

Diagram 6.1
The use of dendrochronology, the
measurement of tree rings, to date the age of wood.

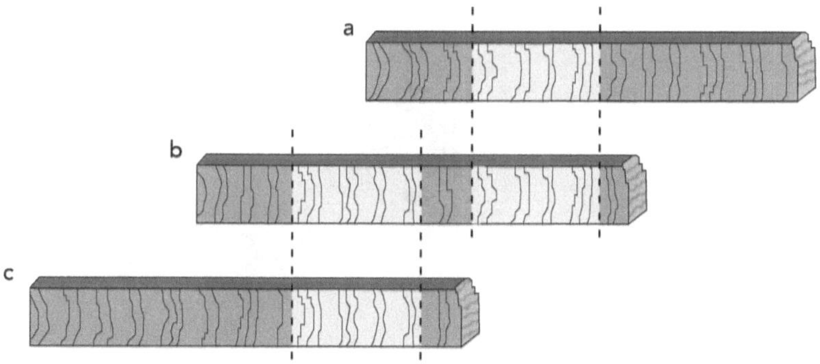

a = Newest wood, newest rings
b = Intermediate wood, providing overlaps back in time
c = Oldest wood, oldest rings

Diagram 6.2
A depiction of a carbon atom, with six protons, six neutrons, and six elections.

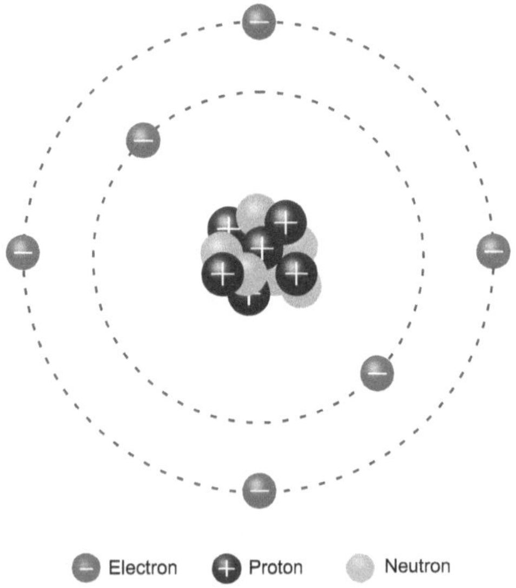

Neutrons are a bit trickier than protons. Many atoms contain an equal number of protons and neutrons. The sum of protons and neutrons comprises the *atomic weight* of an atom. Thus, a standard carbon atom with 6 protons and 6 neutrons has an atomic number of 6 but an atomic weight of 12, also known as Carbon-12 (often written as 12C). Some carbon atoms, however, contain an unequal number of protons and neutrons. One with an additional neutron - that is, 6 protons and 7 neutrons - is Carbon-13 (13C). It is still a carbon atom, but a heavier variety of carbon. A carbon atom with two additional neutrons, 6 protons and 8 neutrons, is Carbon-14 (14C). Carbon-12, Carbon-13, and Carbon-14 are three isotopes (forms) of the same element, Carbon.

We now come to the characteristic of atoms salient to our story. Some isotopes are unstable, and Carbon-14 is one such isotope. This means the atom can change into another atom, a process referred to as radioactive decay. We refer to Carbon-14, in turn, as a radioactive isotope. Let's follow the arithmetic. Carbon-14 has 6 protons and 8 neutrons. Carbon-14 is unstable because a neutron periodically decays into a proton via a process well-documented in particle physics. When this happens, the atom now has 7 protons and 7 neutrons, and it becomes the stable element Nitrogen-14 (14N). It now is no longer a carbon atom. Diagram 6.3 depicts this event.

Ninety-nine percent of the carbon on the earth is Carbon-12. Approximately 1% is Carbon-13, and a rare amount is Carbon-14. Carbon moves into plants via photosynthesis, and into animals through their consumption of plants. Thus, all plant and animal bodies contain a mixture of stable Carbon-12 and Carbon-13, and unstable Carbon-14. When a plant or animal dies, it ceases taking in carbon, and the amounts and relative ratios of Carbon-12, Carbon-13, and Carbon-14 in its body become fixed. In radiocarbon dating, a dating technique using Carbon-14, this is referred to as *zeroing the clock*.[156]

Radioactive isotopes, including Carbon-14, decay at a regular rate. The rate of decay is termed the half-life of the element. The half-life is the amount of time it takes for one-half of a sample (half of all the atoms in the sample) to decay. It is not possible to predict

Diagram 6.3
The decay of Carbon-14 into Nitrogen-14.

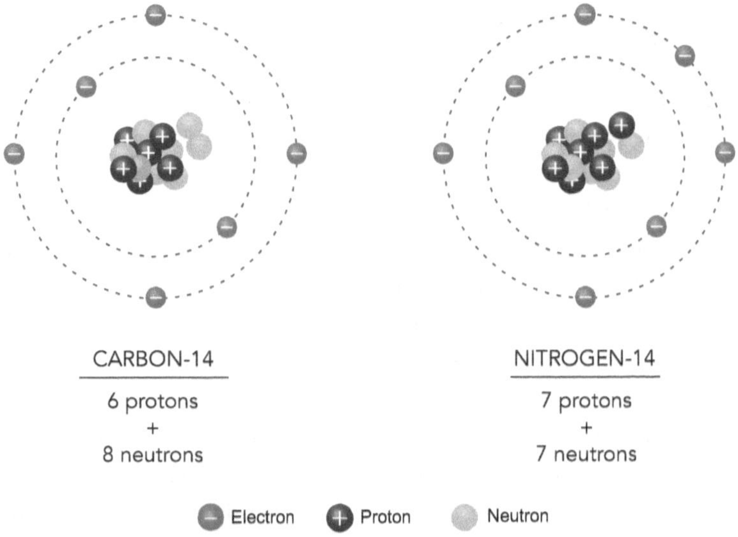

when a particular atom will decay, but because the rate of decay is constant, one can predict with remarkable accuracy how long it will take for half of the atoms in a sample to decay.

The half-life of Carbon-14 is 5,730 years. This means that after 5,730 years, half the Carbon-14 atoms in a sample will have decayed into Nitrogen-14. By measuring the relative proportions of stable carbon to the remaining Carbon-14, the age of an object can be determined.

Because all living organisms contain carbon, radiocarbon dating can be applied to a wide variety of samples: bones, wood, and plant fibers, as well as products crafted from once-living organisms, such as cloth and leather. The technique can be used to date such objects back to around 50,000 - 60,000 years. This simply is a function of the half-life of Carbon-14. As mentioned, after 5,730 years half of a sample of Carbon-14 will have decayed. After another generation of 5,730 years, another half of the remaining Carbon-14 will have decayed, leaving one-fourth of the original amount. After a third generation of 5,730 years, yet another half of the remaining Carbon-14 will have decayed, leaving one-eighth of the original amount, and so on. After several half-life generations, insufficient Carbon-14 remains in the sample to reliably date the object.

It should be noted that radiocarbon dating requires specialized equipment and careful execution. Incorrect results can stem from several factors, including variation of the ratio of non-radioactive carbon to Carbon-14 in the atmosphere; the variation of the same ratio by hemisphere due to carbon exchange with the ocean; the acquisition of carbon in freshwater from rocks; the different rates of carbon take-up in some plants and animals; and the contamination of samples by other carbon sources.

The risk of dating errors, however, can be mitigated. One method is to use an object of known date, such as a book known to have been created in the year 1200 CE, for example, to calibrate the radiocarbon date. Dendrochronology can also be used as a calibration reference. And, as with dendrochronology, a range of samples dated by independent laboratories heightens the reliability of radiocarbon dating. Note that radiocarbon dates typically are expressed in ranges. Thus, an object may be said to be 9,300 - 9,700 years old.

As an aside, both radiocarbon dating and dendrochronology can be used to date objects that do not contain carbon. When a buried cache of such objects (coins) is discovered near carbon-containing objects (a leather bag), the latter can be used to date the former, on the assumption all objects were buried at the same time.

Carbon is but one of several elements whose isotopes permit the dating of objects. Perhaps surprisingly, through the use of a more general technique known as radiometric dating, isotopes of other elements can be used to date rocks. Here, in brief, is the story. Geologists separate rocks into three types: igneous, sedimentary, and metamorphic. Igneous rocks form from the cooling of magma and lava. Sedimentary rocks form from particles of sand, pebbles, and other materials that are deposited together, and then compacted, most commonly on or near the earth's surface. Metamorphic rocks form when existing rocks, either igneous or sedimentary, are transformed by the intense temperatures and pressures under the surface of the earth.

Radiometric dating can be performed only on igneous rock, and only once the magma has cooled. Above certain temperatures, the radioactive decay of any isotope is lost - it diffuses from the magma. As the magma cools, however, subsequent radioactive decay is trapped within the rock. It is this process that *zeros the clock* in igneous rocks.[157] Examples of elements that can be used for dating, as well as their half-lives, appear in Diagram 6.4.[158]

Radiometric dating has enabled scientists to date a wide assortment of igneous rocks. Some of the oldest come from the Jack Hills in Western Australia where zircon, a mineral, is found in the rocks. Dating the zircons indicates the rocks are roughly 4.4 billion years old. Dating igneous rocks at other locations worldwide corroborates this figure, indicating the earth itself is about 4.5 billion years old.[159] Here again, the obligatory caveat applies. Radiometric dating produces its most reliable results when a considerable number of ig-

Diagram 6.4
The half-lives of various radioactive isotopes.

Isotope	Half-Life
Carbon-14	5,730 Years
Aluminum-26	740,000 Years
Iodine-129	17 Million Years
Samarium-147	108 Million Years
Uranium-235	704 Million Years
Potassium-40	1.3 Billion Years
Uranium-238	4.5 Billion Years
Thorium-232	14.0 Billion Years
Rhenium-187	41.6 Billion Years
Rubidium-87	49.0 Billion Years

neous rock samples are examined from diverse locations around the globe, each tested by multiple and independent laboratories.

Techniques to date the age of the sun and moon yield similar numbers, denoting all three objects formed around the same time. These techniques include radiometric dating of rocks returned from NASA moon missions; radiometric dating of meteorites; analysis of crater accumulation on the moon (which, over long timescales, is thought to occur at a constant rate); analysis of changes in the mass and brightness of the sun; measurements of nuclear fusion (the rate at which hydrogen fuses into helium in the core of the sun); and computer simulations of stellar evolution.

Yet another technique to date the age of objects is luminescence dating. Various forms exist, but a general description is as follows. Minerals such as quartz and feldspar exhibit a crystal structure - a lattice - with their atoms aligned in an ordered arrangement. As the atoms absorb energy from various types of radioactive decay, electrons separate from their nuclei but remain trapped within the lattice. Over time, more and more trapped electrons build up within the lattice. Exposure to sunlight releases the confined electrons. Thus, the burying of a sample and its consequent shielding from sunlight zeros the clock. When a buried sample is recovered and exposed to light in a laboratory, the trapped electrons escape and release energy in the form of emitted light - a luminescence. Measurement of the quantity of emitted light determines the age of the sample.

In the case of radiocarbon dating, we noted the technique can date non-carbon containing items (coins) that were buried with carbon-containing objects (a leather bag). The same is true with luminescence dating. The technique can date objects containing crystalline minerals, such as pottery. It can similarly estimate the age of other objects that do not contain, but were buried with, crystalline minerals.

Luminescence dating is referred to as an "intermediate age" technique, as it can measure the age of objects up to 300,000 years old. This might prompt you to wonder: If the technique cannot reach the timescales required to estimate the age of the earth, why is it relevant to our narrative? For three reasons. First and foremost, the technique is capable of refuting all young-Earth hypotheses. Suppose a hypothesis states earth is 10,000 years old. If luminescence dating reliably dates an object to, say, 200,000 years, the hypothesis is refuted. Second, it provides an independent method to calibrate dendrochronology and radiocarbon dating, as well as the age of objects measured by those techniques. And third, luminescence dating assists in measuring the age of fossils, a topic we explore in greater detail in the next chapter.

Reason's journey to determine the age of the earth was considerably lengthier than described so far, attended by many false steps along the way. An early method of estimating the earth's age was first to estimate the life of the sun. Creative thinkers suggested the sun was burning wood, coal, or similar substances. German physicist Hermann Ludwig Ferdinand von Helmholtz (1821-1894) was one of the individuals who disproved these ideas. He was able to show that if the sun derived its energy from coal (or any chemical source), it would have a life expectancy of only 5,000 years. Helmholtz suggested the source of the sun's energy was gravitational contraction and calculated an age of 20 million years.

The Scottish-Irish physicist William Thomson, Lord Kelvin (1824-1907), also explored gravitational contraction, while further suggesting the sun's heat might derive from the periodic impact of meteors falling upon its surface. Informed by these ideas, Kelvin calculated the age of the sun at between 20 million and 60 million years.[160] Kelvin subsequently calculated the age of the earth based on its cooling rate - the time required for the planet to coalescence from a molten sphere into a solid orb. This approach led Kelvin to estimate the earth solidified sometime between 20 million and 200 million years ago.[161] These and other ideas arose before the age of nuclear physics.

Others based their estimates on phenomena closer to home. The Italian geologist Giovanni Arduino (1714-1795) studied the geology of northern Italy and proposed the first classification of geologic time, suggesting the earth was older than Biblical accounts. The

French naturalist Georges-Louis Leclerc, Comte de Buffon (1707-1788), produced an estimate of 75,000 years based on the rates of cooling metals. Further north, Scottish physician James Hutton (1726-1797), English geologist William Smith (1769-1839), and Scottish geologist Sir Charles Lyell (1797-1875) each suggested the earth was millions of years old.[162] They independently based their views on observed rates of erosion, sedimentation, rock layering, volcanism, and other slow-moving natural processes. Lyell believed the earth was more than 300 million years old. The English astronomer Edmund Halley (1656-1742) proposed estimating the age of the earth by measuring the accretion of salt in the ocean. The idea, followed up in the nineteenth century, suggested an age of 200 million years.[163]

With the advent of particle physics at the dawn of the twentieth century, Reason recast its estimates of the earth's age. Discoveries revealing the structure and behavior of atoms spawned the aforementioned technique of radiometric dating. The American scientist Bertram Boltwood[164] (1870-1927) and British geologist Arthur Holmes[165] (1890-1965) were the first to demonstrate the technique based on the decay of uranium into lead. Their initial analysis of rock samples yielded dates from 400 million to 2 billion years old, suggesting a radically older age for the planet than previously surmised. Over the next decades, radiometric techniques improved, and researchers evaluated a wider range of rock types. By mid-century, scientists calculated the age of the oldest rocks, and hence the earth itself, at 4.5 billion years, the currently accepted figure.

While Reason was busy determining the age of the earth, Religion produced its own estimates. Let's return to William Ussher, whom we met at the beginning of the chapter. The Protestant Archbishop calculated the earth was created in 4004 BCE, but he was not alone in his estimate. The Turin King List, an Egyptian papyrus dating from approximately 1550 BCE, refers to a reign of the gods beginning in 39,670 BCE. Other Egyptian sources provide a date of 17,680 BCE, correlating to the reign of various deities and demigods. Zoroastrianism provides a date, resting solely on the Prophet Zoroaster, of 12,000 BCE.

Early Judaic sources (the Seder Olam Rabbah, the Seder Olam Zutta) suggest dates around 4000 BCE. Later Judaic and Christian sources (the Septuagint, a translation of the Hebrew Bible; the Masoretic Text) suggest dates ranging from 4000 to 5500 BCE. Eusebius of Caesarea (c.260-c.339), a Greek historian of Christianity, records a date of 5228 BCE. The Christian monk, Maximus the Confessor (c.580-662) calculated a date of 5493 BCE. Moses ben Maimon (c.1135-1204), the Andalusian Jewish philosopher, calculated 4058 BCE.[166] The list goes on.

Religion commonly arrives at estimates of a younger earth because it relies on the prophecy of one individual or the literal interpretation of a consecrated text. Different prophets, translations, and interpretations produce different dates. A common technique, for instance, is to count Biblical genealogies, calculating back in time from *begat* to *begat*. Another is to apply rates of reproduction, counting the number of years for the global population to increase from the first two humans, Adam and Eve, to its contemporary level. Another approach is to start counting from a mythical event, such as the flood discussed in *The Origin of Religion*. The trick here, of course, is first to invent a date for the Biblical inundation. Allot credit to Hinduism, however, the only religion to provide a date on a similar timescale as Reason. Hinduism, with its cycles of death and rebirth, dates the earth at 8.6 billion years old.[167]

The age of the earth is not the only time-based event where Religion has expressed its opinion. The development of the modern calendar provides further insights into how Religion and Reason operate. In 1267 the English Franciscan friar and philosopher, Roger Bacon (c.1219-c.1292), sent a missive to Pope Clement IV, informing him the Julian calendar year was several minutes longer than the solar year. This discrepancy, he said, would accumulate year by year, leading Christians to celebrate their holy days on the wrong dates.

To redress the issue, Bacon advised the pope to reform the calendar, a brave act in an era when those who challenged the Catholic Church risked severe penalties.

Fortunately for Bacon, his appeal was received by a lawyer named Guy Le Gros Foulques, who previously was in contact with Bacon. In 1266 Le Gros Foulques was elected Pope Clement IV. Yet unfortunately for Bacon, Pope Clement IV died in 1268, having taken no action on the calendar. Even more unfortunate for Bacon, he lost his protector. A novel thinker who emphasized the study of nature by empirical methods, Bacon later was charged with teaching "suspected novelties," was convicted, and imprisoned.[168]

Before Bacon, others also noticed the discrepancy in the calendar year and sought to rectify it. One was a monk named Dionysius Exiguus (c.470-c.544), whose motive was to ensure that Easter was celebrated on the correct date. In deference to the piety of his era, he publicly declared the date should be inspired by the Holy Spirit. After which he privately employed mathematics and astronomy to calculate the dates for Easter for the next 95 years.[169]

Further east, Muslim scholars also sought to bolster the precision of the Islāmic lunar calendar. During the Golden Age of Islām, construction of observatories and improvements in scientific equipment (astrolabes, sundials, and globes) led to more accurate astronomical observations.[170] As in the West, the impetus was religious. A more accurate calendar would permit Muslims to better observe holy events, such as Ramadān, the ninth month of the Islāmic calendar, throughout which followers fast during daylight hours. Similarly, improved measurements of the angle of the sun would provide more accurate *salāt* times, the five times of day for the religious duty of prayer. Islāmic scholars succeeded in their mission, producing a calendar said to be accurate to the orbit of the moon within one day every 2,500 years. As an aside, the astronomical observatories of the Golden Age also enabled Muslim scholars to arrive at many of the same discoveries mentioned in *The Case of Gravity*.

Back in the West, three centuries after the appeal from Bacon, Pope Gregory XIII (1502-1585) appointed a commission to correct the calendar.[171] The commission completed its work in 1581. One year later, in 1582, Gregory signed a bull marking the last year of the Julian calendar, the calendar established by Julius Caesar in 45 BCE. The new Gregorian calendar was introduced on 5 October 1582. To address the discrepancies in the Julian calendar, a 10-day adjustment was immediately implemented, meaning the day after 4 October 1582 was 15 October 1582. The new calendar was gradually adopted throughout Europe and across the globe: in 1752 in England, in 1753 in Sweden, in 1873 in Japan, in 1912 in China, in 1918 in Russia, and in 1924 in Greece. It remains the most widely-used calendar today. Perhaps Pope Gregory XIII should be congratulated for achieving a complex and bureaucratic reform of the calendar. Alas, he also appointed a committee to update the Catholic Church's *Index of Prohibited Books*.

I mention the story of the calendar to modify an earlier comment from chapter three. There, I used the term *natural philosopher* rather than *scientist* for the early pioneers who undertook the philosophical study of nature and the physical world. I now differentiate between *natural philosophers* and *natural theologians*. Both practice a form of proto-science to gain new knowledge, but from different motivations. The former search for new knowledge and objective truth, following wherever evidence might lead. The latter, however, merely seek to ratify their deity and their existing religious beliefs. The hypotheses of natural theologians frequently are unsound, as they typically ignore the first criterion articulated by William Whewell in *The Origin of Reason*: that a hypothesis should predict novel phenomena. Moreover, they rarely formulate hypotheses to be falsifiable. The religious inducements to improve the Gregorian and Islāmic calendars led to admirable results for Religion and Reason alike. Yet, in other instances, natural theologians often are constrained by the questions they are willing to ask and the results they are willing to accept.

Natural theology explains why Religion routinely cites erroneous dates for the age of the earth. An individual who believes scripture is the sacrosanct, inerrant word of a deity will seek answers that fit safely within the dogma. They will devise interpretations that reinforce doctrinal teachings. The believer resists looking outside, for fear of finding explanations that contradict the dogma. He might even invent explanations, merely to adhere to doctrinal truth. This is a common trap natural theologians fall into. The Danish geologist, Nicholas Steno (1638-1686), demonstrates the point. Trained on classical scientific texts, Steno conducted early work on fossils and rock layering. He later converted to Catholicism, abandoned his geological work, became a priest, then safely concluded the earth was 6,000 years old based on a Biblical framework.[172] The English theologian Thomas Burnett (c.1635-1715), arguing from Scripture, also calculated the earth to be some 6,000 years old.[173]

Allow me to linger a bit longer on Burnett. In the late seventeenth century he attained a position of prominence as chaplain to William of Orange, the king of England, Ireland, Scotland, and parts of present-day Netherlands. Burnett leveraged his education and credentials to author several works, the most famous of which was *The Sacred Theory of the Earth*, penned in 1681. The text investigates the Biblical flood cited in Genesis. Burnett accepts the authenticity of the deluge but, uncommonly, endeavors to explain salient features of the event. He begins by chronicling the flood as a global event, not a regional or local phenomenon. He then asks how much water is required to inundate the planet. He considers the heights of mountains, the depths of oceans, and the portion of the planet already enveloped by water. The quantity of water required to submerge all land, he concludes, is equal to eight times the size of the world's oceans. Burnett does not explicitly depict his calculation, but several assumptions allow us to emulate the arithmetic.

Let's assume the tallest mountains extend eight kilometers above sea level (Burnett assumed one mile). Our premise is not unreasonable, as Mount Everest reaches nearly nine kilometers. We will assume the ocean is two kilometers deep (Burnett assumed one-quarter mile) and covers one-half of the globe (Burnett also assumed one-half). We will use 6,400 kilometers as the radius of the earth, the same number derived in *The Case of Gravity*. Our calculation, expressed as a set of concentric spheres, appears in Diagram

Diagram 6.5

A depiction of Burnett's calculation of required flood waters.

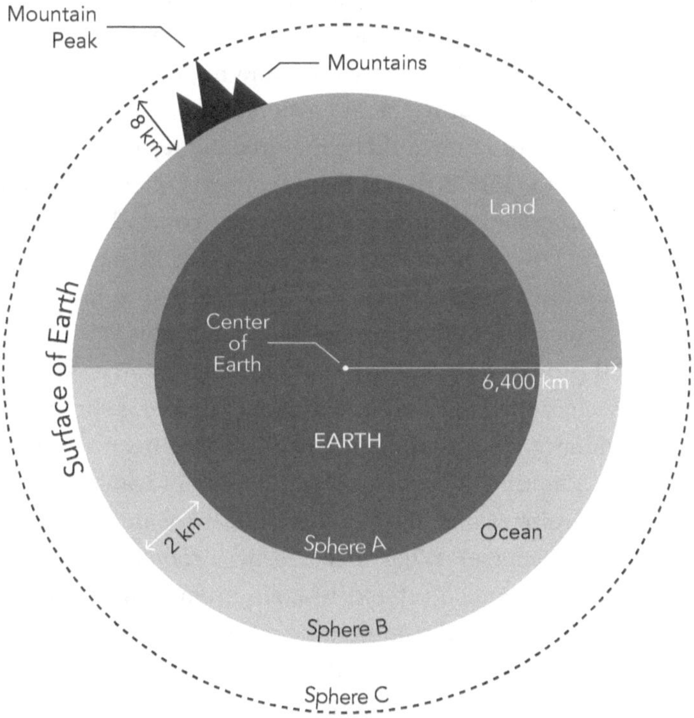

THE MATH

Volume of Spheres

$A = 4\pi / 3 * (6{,}398 \text{ km})^3 = 1{,}096.5 \text{ billion km}^3$

$B = 4\pi / 3 * (6{,}400 \text{ km})^3 = 1{,}097.5 \text{ billion km}^3$

$C = 4\pi / 3 * (6{,}408 \text{ km})^3 = 1{,}101.6 \text{ billion km}^3$

\# of oceans = [Total water required] / [Size of existing ocean]

= [C - B] / [(B - A) / 2]

= 4.1 billion km^3 / 0.5 billion km^3

= 8

6.5. Applying geometry to compute the volume of each sphere produces the same result as Burnett. A quantity of water equivalent to eight oceans is required to inundate the planet.[174]

Having determined the necessary quantity of water, Burnett turns his attention to its source. From where did the water spring? Burnett takes into account moisture in the air, precipitation in the clouds, discharge from rivers, and possible celestial sources. Interestingly, he calculates that "forty days and nights" of rain, as cited in the Bible[175], are insufficient to create eight oceans of water. He suggests 45 years of rain would be required but rejects this possibility as contrary to Holy Writ. He likewise rejects other atmospheric, terrestrial, and celestial sources of the deluge, concluding the water resides in the bowels of the earth. According to Burnett, the eight oceans, stored in a hollow earth, rise when needed to inundate the planet, then recede into the abyss as the flood abates.[176]

Burnett provides a beautiful example of natural theology - contriving a scenario to ratify existing dogma without a shred of supporting evidence. To his credit, Burnett initiated his investigations with sound scientific questions. What quantity of water is required to flood the entire planet? From where did the water emanate? To where did it disperse? What mechanisms propelled the deluge? Burnett's assertions also offered testable predictions. Is the earth hollow? Is there a vast chasm holding a repository of water below the surface of the earth, awaiting the next global inundation?

Elsewhere in his text, Burnett calculated the date of the flood (1600 BCE, from his reckoning) by extrapolating in reverse. Starting from the population level of his era, he computed back to the time that Noah and his family exited the ark - a sound approach, at least in principle. Burnett clearly was no slouch. Yet he was able to interpret data and formulate conclusions only in context of his religious beliefs. In an era where Religion offered the only widespread paradigm for imparting truth, he was unable to objectively assess the evidence or embrace an impartial verdict: The flood story is a myth.

I earlier referred to Burnett's claim that the earth is less than 6,000 years old. That is not quite correct. To quote him precisely from chapter two of his opus: "It is now more than Five Thousand

Years since our World was made."[177] Let's commend Burnett's foresight for inserting the word *more*. The earth is indeed *more* than 5,000 years old. It is 900,000 times older.

OBSERVE

In its quest to determine the age of the earth, Reason enlisted all five steps from its primary *how to think* process, the scientific method. From *The Case of Earth's Age,* however, we earlier promised to elaborate on the third and fourth steps, Observe and Verify.

Observe puts hypotheses to the test. Experiments are conducted, observations are recorded, outcomes are dispassionately analyzed. A meticulous appraisal begets the moment of truth when a hypothesis is either confirmed or refuted. Apart from stellar simulation, which is an explanatory and confirmatory technique, all of Reason's dating techniques mentioned in this chapter typify the practices of Observe. Direct and indirect observations were logged of geological strata, rock types, tree ring patterns, decay rates of atoms, by-products of elements, the moon's craters, the sun's mass, and much more. Scrupulous assessment of the results led to acceptance or rejection of specific dating hypotheses.

Religion possesses no comparable process to Observe. It lacks the ability to objectively design experiments, sedulously collect empirical evidence, and impartially assay the results. It also is unwilling to subject its claims to possible refutation. Undeterred, Religion employs natural theology and other techniques from the five sacreds to mimic Observe. Yet its attempts routinely fall afoul of logical fallacies. Let's explore several common instances.

APRIORISM Apriorism is the fallacy of refusing to view evidence that might contradict one's beliefs. We witnessed one instance in *The Case of Gravity,* when the Italian philosopher Cesare Cremonini (1550-1631) refused to look through Galileo's telescope. He chose not to see moons in orbit around Jupiter, which would belie the geocentric teachings of the Catholic Church. Apriorism allows Religion to imitate Observe by selecting only favorable evidence.

POST HOC FALLACY This fallacy takes the form *if B follows A, then B must have been caused by A*. The conclusion derives from the order of events rather than their causality. Prayer illustrates the fallacy. A petitioner who subsequently experiences a favorable outcome often credits the power of prayer. Consider the following: "I really wanted that new job. I worked hard, and I prayed about it. Two months later my boss called me in, and guess what, I got the job!" Believers routinely cite similar "evidence" to affirm the efficacy of prayer. It is, if nothing else, a studied ruse, purposefully self-reinforcing. Yet Reason recognizes the claim as a *post hoc ergo propter hoc* fallacy. Sequence, devoid of substantiating evidence, cannot assert cause. Religion, true to form, offers no evidence of causality.

APPEAL TO FORCE The protagonist of this fallacy threatens an undesirable outcome if a claim is not accepted. He relies on fear and intimidation rather than evidence to assert a point of view. Theologians regularly threaten ostracism, damnation, and eternal perdition to affirm their teachings. Sacred texts endorse the threats. The Bible, for example, admonishes, "The wicked shall be turned into hell, and all the nations that forgot God."[178] The Qur'ān fulminates, "Allāh will gather the hypocrites and disbelievers in Hell all together."[179] The Gurū Granth Sāhib forewarns, "Those who practice falsehood are dogs; those who slander the Guru shall burn in their own fire. They wander lost and confused, deceived by doubt, suffering in terrible pain. The Messenger of Death shall beat them to a pulp."[180] When transgressors suffer, Religion then selectively interprets their fates as attestations of divine will: The Roman Empire succumbed from its perverse pagan traditions; the sins of Lisbon's inhabitants engendered the cataclysmic earthquake of 1755;[181] and homosexuality begat the curse of AIDS. Religion adopts the selective interpretations as evidence of its claims.

VERIFY

We now turn to Verify which, as you recall, is the elevation of a broad set of confirmed hypotheses to an accepted theory. For the theory to be accepted, it must be independently verifiable. The theory that the earth is 4.5 billion years old satisfied the verification challenge by obtaining and examining large numbers of samples from around the world, by analyzing samples in independent laboratories operated by independent scientists, and by cross-checking and validating the results of disparate dating techniques. Independent verification also played a role in Reason's earlier, incorrect estimates of the earth's age, by weeding out erroneous hypotheses - an exemplar of the self-correcting nature of Reason.

Religion occasionally purports to embrace independent verification. Yet closer scrutiny frequently discloses a symbiosis. The claimant and the verifier not only embrace the same faith, but nourish a shared craving to ratify their faith - a creative but diametric interpretation of independence.

Faith-based verifications, of course, play no role in Reason. Its *how to think* process transcends faith. If you doubt this fact, consider this point. In Reason, a Muslim can verify a discovery by a Sikh, an atheist can corroborate a finding by a Jew, and a Taoist can confirm a claim by a Christian. The truths of Reason thus impart a universal veracity wholly unadulterated by religious affiliation, political inclination, or cultural sentiment. Religious apologists can assert a similar claim only with chagrin.

The reality is that Religion is unable to leverage the potency of Verify. It is unwilling to forsake dogma in light of new evidence, and it lacks independent verification of its claims. These flaws are endemic. Religion contains few of the checks and balances vital to self-correction. Religion resists deep introspection, meddlesome inquiry, and outside criticism, preferring instead to reside comfortably within its dogmatic echo chamber. Once "bad" information enters, it is difficult to evict. Religion's ersatz attempt at Verify once again relies on the deliberate use of logical fallacies. Let's cite three further examples.

CIRCULAR REASONING Circular reasoning declares a premise valid, then presents a conclusion as a restated form of the premise. The fallacy takes the form *A is true because A is true*. Religion embraces the fallacy when it declares scripture to be the true word of a deity, then bases its claims on the declared truth of scripture. This translates into a self-verifying *scripture is true because scripture is true* argument. Circular reasoning arguably is the most common fallacy employed by Religion, serving as the foundation for myriad dogmatic claims. It also is the basis for Religion's erroneous estimates for the age of the earth which, in the minds of the steadfast, must align to scripture. Religion's insistence on the truth of Holy Writ warrants another comment. Apart from circular reasoning, consider that there is no sentence in any religious text that could not have been written by an author born in the prevailing age. There is no pronouncement, no affirmation, no revelation that discloses knowledge unavailable from that age. There is no equation, graph, or calculation that conveys a robust, mathematical comprehension of phenomena. Consider, further, a declaration from the Qur'ān: "And it was not [possible] for this Qur'ān to be produced by other than Allāh."[182] Reason explicitly denies this claim. Any number of creative, talented storytellers could have penned a religious text, their writings constrained only by the murky knowledge of their era.

FALSE ALTERNATIVES This is a double fallacy. To begin with, the advocate falsely asserts a limited number of alternatives. He then contends that one, and only one, of the limited alternatives is true. A dirigiste proclamation from theologians, quoting words attributed to Jesus, illustrates the fallacy: "He that is not with me is against me."[183] The assertion is phrased to admit no third possibility, no degree of gradation, no nuance. The advocate, implicitly or explicitly, then professes "with me" as the correct choice. There is no discernment among elections, no evidence to support any alternatives, and no acknowledgment of neutral, agnostic positions. In the context of Verify, the advocate "verifies" the claim by creating a false dichotomy and by suppressing viable options.

APPEAL TO THE GALLERY Proponents of this fallacy attempt to verify the truth of a claim by appealing to the emotions, passions, and popular opinions of their audience. Secular appeals might emphasize patriotism or political affiliations. Religious appeals stress love, guilt, fear, or salvation. Some assert the fallacy is most successful with undiscriminating audiences; others affirm we all are susceptible. Regardless, it is an effective and (sometimes) legitimate method of persuasion when supported by compelling evidence, but invalid when solely appealing to established beliefs and prejudices. Mother Teresa's stance on divorce illuminates the fallacy. In 1995 the citizens of Ireland voted on a constitutional amendment to legalize divorce. Mother Teresa (1910-1997), a Saint of the Catholic Church, opposed the referendum, stating her religious and secular objections in a letter to the Irish people. Divorce is contrary to the teachings of Jesus;[184] the family that prays together stays together; divorce harms children and society; divorce creates temptations.[185] Her missive appealed to popularly-held principles of faith, love, and marriage in Irish society to verify her position. Yet she offered no evidence, and she acknowledged no equally relevant grounds for the dissolution of marriage (physical abuse, emotional abuse, insanity, abandonment, or addictive behaviors). Reason recognizes the irrelevancy of gallery appeals when unattended by sound, substantiating evidence. So did most Irish. Of 1.6 million votes cast, the referendum passed by 9,118 votes.

Logic, the field that studies the form and validity of arguments, classifies numerous types of false reasoning. The number varies by source. Damer, for example, catalogs 64 types.[186] A full treatment of logical fallacies is beyond the scope of this book. Yet as the examples in this section suggest, Religion borrows heavily from the catalog, leveraging the fallacies to articulate its claims and spread its dogma.

A final example further illustrates the point - an examination of Pascal's Wager, arguably one of the more notable instances of logical fallacy.

The wager appears in *Pensées* (*Thoughts*), an unfinished and posthumously published work by the French mathematician and philosopher Blaise Pascal (1623-1662). Pascal frames the wager as follows: Suppose you must choose to either affirm or deny the existence of god. If you choose to disavow his existence, you gain a finite set of worldly pleasures - the latitude to participate in proscribed pursuits (drinking, gambling) free from spiritual guilt. But if you choose to affirm his reality, "There is here an infinite life of infinite happiness to gain,"[187] both in the temporal and spiritual realms. Since the possible reward is incomparably greater than the potential loss, Pascal argues that the wise choice is to wager on the existence of the deity, to accept Holy Writ, and live one's life accordingly.

A cursory examination of the wager might suggest a well-grounded, common sense validity. A circumspect assessment, however, identifies multiple logical fallacies. We will mention three. To begin, the observant reader will note that Pascal's Wager, by offering only two incompatible alternatives, suffers from the previously described fallacy of *false alternatives*. There is no third path. There are no shades of gray.

Second, the wager leverages *appeal to tradition*, a logical fallacy that stresses reverence for traditional values and ways of life to validate an argument. The tradition in Pascal's Wager is Christianity. Pascal, a converted and devout believer, centered his wager on a Christian god. It is no surprise Pascal failed to address this fallacy. The wager resonated among a population reared and catechized under Catholicism - a people harboring respect and reverence for Christian institutions, practices, and beliefs. The wager thus exploits regard for Christian teachings to extend an existing paradigm. Reason recognizes that traditions often originate from sound causes, and that they perform a prominent function within the social fabric of a community. Yet respect for traditions, and the traditions themselves, are irrelevant indicators of an argument's validity.

The English philosopher John Stuart Mill (1806-1873) penned one of the more notable critiques of appeals to tradition: "The despotism of custom is everywhere the standing hindrance to human advancement."[188]

Third, the wager embraces the *question begging definition* fallacy, which attempts to establish an irrefutable argument predicated on the definition of a key term. The term, rather than the argument itself, becomes the non-falsifiable basis of the claim. In the case of Pascal's Wager, the definitional term is *god*. If one accepts the wager, why is a Christian god the correct god? Why is it not better to wager on a Sikh or Shintō god? And why should one wager on only one god? To obtain the greatest possible chance of reward, would it not be better to acknowledge all possible gods - animist, polytheist, and monotheist? To demonstrate one's commitment to each deity, why not observe the Muslim Sabbath (Friday), the Jewish Sabbath (Saturday), the Christian Sabbath (Sunday), and the Buddhist Sabbath (the day of the week varies by the lunar calendar)? Why not abstain from beef (Hinduism), pork (Islām), lobster (Judaism), coffee (Mormonism), and potatoes (Jainism), living one's life in accordance with the forbidden foods from each religion? Pascal not only posits his deity as a non-falsifiable premise, but he fails, unsurprisingly, to explain why the wager pertains only to his preferred deity.

As noted, Religion's labored relationship with logic extends beyond Observe and Verify, encompassing the full spectrum of its dogmatic claims. This is inevitable. Its inability and unwillingness to produce rigorous, impartially evaluated evidence compels Religion to rely on *a priori* reasoning (defined, in this instance, as reasoning from wishful propositions, absent of experience). Equally unable and unwilling to subject its pronouncements to independent verification, Religion invariably falls afoul of *non sequiturs*.

Reason, too, periodically stumbles into the trap of logical fallacy, but with a marked distinction. Reason attempts to identify and correct false reasoning. It elevates testable predictions, supported by empirical evidence, above the cut and thrust of deductive argument. Reason understands that the best-reasoned arguments collapse in the face of nullifying empirical evidence. Religion, rejecting this reality, consigns itself to a dominion of illogical, unsubstantiated, and continually disproved claims.

The Case of Earth's Age articulated the approaches employed by Reason and Religion to estimate the age of the earth. From its first serious estimates in the eighteenth century, Reason formulated, tested, accepted, and discarded various hypotheses over the next two centuries, ultimately adopting 4.5 billion years for our planet's age. Religion, basing its methodology on the five sacreds and notional reasoning, in nearly all instances arrives at a younger age for the planet. Challenges to its estimated dates are countered by Religion's standard response, appeals to authority. Unsurprisingly, modern practitioners of "creation science" (a misnomer bearing no resemblance to genuine science) typically estimate the age of the earth at less than 10,000 years, the same as their historical predecessors, derived from the same artifices of natural theology and casuistry.

Diagram 6.6 presents the conclusions from *The Case of Earth's Age* for the Observe and Verify steps of the scientific method.

MISCELLANY

This chapter refers to 92 naturally occurring elements. The actual number ranges from 88 to 98, varying by source and definition. The higher number includes a handful of trace elements.

Unstable Carbon-14 decays into stable Nitrogen-14 via a form of beta decay, where one of the neutrons in the Carbon-14 atom decays into a slightly less massive proton, simultaneously emitting two other subatomic particles (in the jargon, an electron and an electron antineutrino).

It is not quite true that radiometric dating can be performed only on layers of igneous rock. Sedimentary layers that contain granules of igneous rock also can be dated. The challenge relates to the origin of the various granules. Each grain might have formed in a separate location, then been transported by erosion to the sedimentary layer. Analyzing ten different grains might yield ten different ages.

Hinduism, among its multiple schools and sub-schools, offers a number of creation stories. Many embrace the cyclical creation and destruction of the universe, a cycle that began its course with the birth of Brahma over 100 trillion years ago.

As an aside, the Gregorian calendar maintained 1 January as New Year's Day, but there is no astronomical significance to 1 January as the start of a new year.

In this chapter I am a bit hard on deductive logic, and equally hard on rationalism in so far as it is anchored by deduction. Deduction dangles an all too common temptation to drive specious conclusions from unsound premises. Yet a more thoughtful view acknowledges the role and importance of deduction in advancing human understanding across many disciplines. Deduction accompanied by induction yields amazing truths. But the salient point is this: In cases where intellect-driven, premise-drives-conclusion, *a priori* logic battles the powers of empiricism and testable predictions, their approaches yielding contradictory results, induction reigns victorious. In every case, bar none.

A summary of the different approaches of Religion and Reason in the third and fourth steps of the scientific method, Observe and Verify.

Diagram 7.6

Principles

	Religion	Reason
① ASK	Compels adherence to dogma; suppresses, even punishes, inquiry that challenges existing beliefs and convictions.	Encourages a culture of openness, advocates inquiry that challenges established knowledge.
② PREDICT	Deliberately structures beliefs to avoid testable predictions, posits faith as superior to evidence.	Demands hypotheses with testable predictions that can be confirmed or disproved by observation and experiment.
③ OBSERVE	Selects observations that ratify its beliefs; dismisses, disparages, or condemns contradictory evidence.	Conducts observations and experiments, analyses results to accept or reject hypotheses.
④ VERIFY	Relies on appeals to authority and circular reasoning, professes scripture as truth then argues truth from scripture.	Requires independent verification and a preponderance of evidence; dismisses the authority, prestige, and intelligence of the advocate.
⑤ ACCEPT	To be discussed in Chapter VII.	To be discussed in Chapter VII.

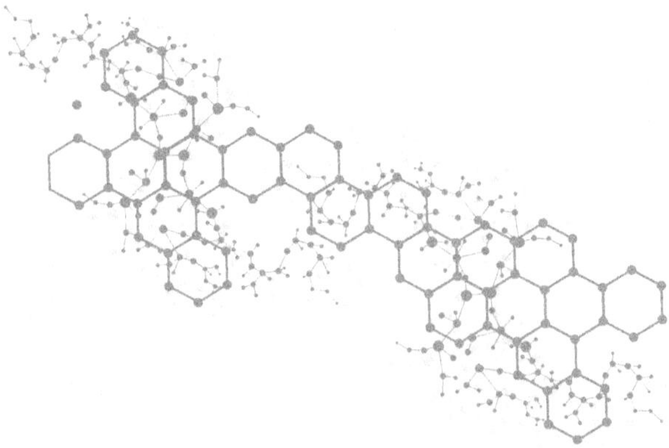

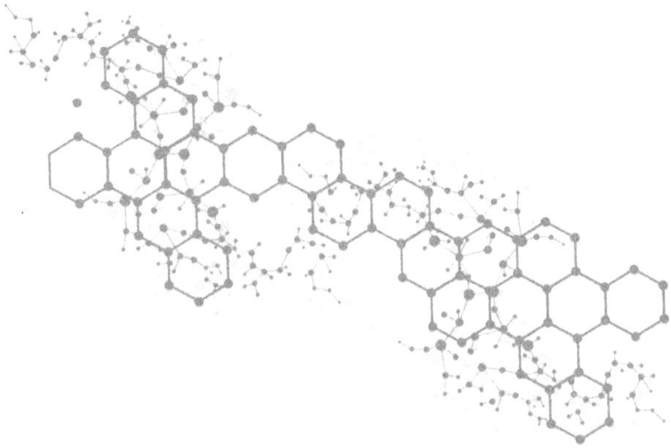

[VII]

The Case Of Human Origins

What is man?
The point I want to make now is that all attempts to answer that question before 1859 are worthless, and we will be better off if we ignore them completely.[189]

GEORGE GAYLORD SIMPSON (1902–1984) AMERICAN PALEONTOLOGIST

A Chinese creation myth contends that life and humanity originated from the fleas in the hair of a giant named Phan Ku.[190] In Iceland, the heat of Muspell melted the ice of Niflheim, causing moisture to descend into the Yawning Gap, from where Imir arose to forge the world and populate it with life. In Japan, two kami stood on the Heavenly Floating Bridge, stirred the waters with a Heavenly Jewelled Spear, allowing drops to fall from the tip and form the land. In West Africa, Olorun, Owner of the Sky, sent a pigeon, a hen, and some soil down to Earth to initiate life. For the Aztecs, the feathered-serpent god Quetzalcoatl created humankind. For the nearby Maya, Huracan, the god of the wind, assisted in the creation of the earth and all life upon it. For the Mongols, the white goose god Kök Tengri and his partner Er Kishi are the creators. For the Iroquois, Sky Woman fell to the earth from a floating island in the welkin to populate the land. In ancient Sumer, the gods Enki and Enlil, along with several of their peers, created the first people. The Brahmanda Purana, a Hindu text, suggests that Brahma emerged from a cosmic golden egg, whereupon he created all the creatures on Earth.[191]

Every major religion offers an account of the creation of life. The method of creation, its location, and the ascribed deities vary, but creation myths share two common elements. First, unsurprisingly, life originates where believers of the myth happen to live. Second, the myths assert the existence of one or more supernatural beings. For most of human history, earthly explanations for the origin of life were rare or non-existent.

With the advent of the Enlightenment and the reemergence of scientific inquiry in the West, the situation began to change. Many no longer believed the creation myths but could offer no viable alternative. Advances in geology, botany, and biology, however, gradually led some to speculate on evolutionary change. The French mathematician Pierre Louis Maupertuis (1698-1759) suggested that of all species ever produced, only a small portion are alive today. The English physician Erasmus Darwin (1731-1802), the grandfather of Charles Darwin, proposed that all life originated from a single organism. The French naturalist Georges-Louis Leclerc (1707-1788) advocated the inheritance of traits from parent to child. The German biologist Ernst Mayr (1904-2005) credits Leclerc for bringing "the idea of evolution into the realm of science."[192] Yet despite these and other efforts, no one was able to discern the mechanism that drives evolution. This changed in 1859 when Charles Darwin (1809-1882) published *On the Origin of Species*, one of the most influential books in the history of humankind.

Our third and final case is *The Case of Human Origins*, where we examine natural selection as the explanation for the rise of humanity. We again consider the respective approaches and conclusions from Reason and Religion. *The Case of Human Origins* also allows us to take a closer look at the fifth and final step of the scientific method, Accept.

Charles Darwin studied medicine at the University of Edinburgh but to the chagrin of his father, he showed no appetite for the profession. He then attended Christ's College, Cambridge, in anticipation of becoming an Anglican parson. While at Cambridge he began collecting and classifying beetles, and he soon developed friendships with several naturalists at the university. One of these friendships led to an opportunity to participate on the voyage that shaped his life. Upon graduation in 1831, he joined the *HMS Beagle* as a naturalist on a five-year trek to South America, New Zealand, and Australia. Throughout the voyage, Darwin collected and described the flora and fauna of various locations, but most notable are his observations from the Galapagos Islands.

The Galapagos is a cluster of 21 islands and numerous islets, located some 1,000 kilometers off the west coast of Ecuador. Darwin noticed that much of the flora and fauna on each island was unique to that specific island. On James Island, of thirty-eight species of plants (now known to exist nowhere else in the world outside of the Galapagos), thirty species grew solely on that one island. On Albemarle Island, twenty-two of twenty-six species of flora were confined to that one isle. Darwin observed the same phenomenon among fauna, noting that iguanas, tortoises, hawks, finches, mockingbirds, and other species differed from island to island. From these observations on species variation, Darwin began to surmise that, in his own words, "one species had been taken and modified for different ends."[193]

Upon returning to England, he continued his research and began to draw parallels with *artificial selection*. Artificial selection is "choosy" breeding, where humans selectively breed plants and animals to cultivate particular traits.[194] Farmers and horticulturists have practiced this technique for millennia. Humans have selectivity bred roses for their beauty and fragrance, sunflowers for the oil from their seeds, apples for their size and taste, cows for greater quantities of milk, sheep for the fineness of their wool, horses for strength and speed, maize for ample cobs and kernels, and dogs for work and companionship. The consequence of thousands of years of artificial selection is that many a creature no longer resembles the species from which it arose in nature. In the parlance of artificial selection, a human is the *choosing agent*, the one making the breeding choices that lead to new traits, typically for purposes of domestication. Darwin's epiphany was that nature also performs the role of choosing agent. This is the process of *natural selection*.

In nature, the choosing agent might be an insect that favors one flower over another from its scent. It might be a female peahen that chooses a male peacock from the attractiveness of its tail feathers. Or it might be a grey squirrel that selects the meatiest acorn. The choosing agents in these examples select a particular target based on its perceived superiority. The selected flower, peacock, and oak tree are more disposed to reproduce, passing their traits to their offspring. It is important to note, however, that a choosing agent in nature does not necessarily imply conscious selection.

Now, take the concept of a choosing agent one step further. Rather than selecting a specific object, it is more common for an agent to select any obtainable object. A tuna chasing a school of tarpon will devour the first one it can catch. A frog will ambush the first fly that comes within range. A lizard will ensnare the first katydid whose adaptive coloring does not match its leafy habitat. In these cases, the faster tarpon, the more cautious fly, and the leaf-colored katydid survive to pass on their traits. Offspring who inherit the traits - generation after generation - come to possess superior survival equipment for their environments. From observations similar in essence, Darwin realized that survival itself becomes the choosing agent.[195]

From human as choosing agent to survival as choosing agent, Darwin further realized that natural selection favors species that are best adapted to their environments. Darwin believed this explained the diversity of life he encountered on the Galapagos Islands. Each island, due to its distance from other landfalls within the chain, offered a unique ecological niche. Natural selection produced species adapted to survive in each environment, resulting in species unique to each island.

If we reverse the clock, natural selection offers another important observation. Each descendant has an ancestor from whom traits were inherited. That ancestor had its ancestor, and so on continuing back through time. Via this chain of antecedents, natural selection implies that all life descends from one or a small number of proto-life forms - a startling claim in the mid-nineteenth century.

With the publication of *On the Origin of Species*, Darwin came to be publicly regarded as the discoverer of natural selection. But to acknowledge credit where it is justly earned, Alfred Russel Wallace (1823-1913) today is credited as co-discoverer.[196] He articulated a similar mechanism of evolution, which was presented to the Linnaean Society of London in 1858, along with an essay from Darwin. Wallace, an autodidact, came to be one of Darwin's staunchest defenders. Today, the Linnean Society awards the suitably-named Darwin-Wallace medal to individuals who make notable contributions to evolutionary biology.

This is a good point to pause and more formally define natural selection as the union of three elements: inheritance, variation, and competition.

> INHERITANCE Offspring inherit traits from their parents. Genes, to be discussed shortly, are the carriers of the traits. Instructions encoded in the genes express the traits.
>
> VARIATION Living organisms replicate themselves but, notably, not with perfect accuracy. This leads to variation in heritable traits among offspring in a population.
>
> COMPETITION With survival as the choosing agent, some traits provide organisms with a superior ability to face the pressures of their environment. The better-adapted organisms, those possessing well-suited machinery for their finite niches, survive and reproduce. Over time, their genes become more widely represented in the population.

The union of inheritance, competition, and variation is evolution by natural selection. To the indiscriminate observer, an organism might appear purposefully designed for a unique environmental niche. Yet natural selection recognizes that organisms only appear purposefully designed. Their biology, their abilities, and their survival are products of trial and error, behind which stands a mindless, goalless mechanism that operates ceaselessly - generation over generation over generation - driving the emergence of new species and the extinction of others. Design, therefore, is a non-intelligent, spendthrift process that, in the words of Dennett, "releases and test-drives many variants, letting the losers die unexamined."[197] Gould provided a pithier description: "The organism proposes and the environment disposes."[198]

Natural selection, by way of explaining adaptive change and the emergence of new species over vast periods of time, provides a workable mechanism for evolution. Combined with the fossil record, it produces a story of life something like this. Life on earth arose approximately 3.9 billion years ago, 600 million years after the formation of the planet. The earliest life forms were bacteria that ruled the earth for 3 billion years. Then, circa 540 million years ago, an event christened the Cambrian Explosion marked the first emergence of modern, multicellular organisms. Fish appeared on the order of 530 million years ago, land plants about 475 million years ago, forests roughly 385 million years ago, mammals around 200 million years ago, flowering plants circa 160 million years ago, hominins approximately 6 million years ago, and our species, *H. sapiens*, some 200,000 years ago. Relative to the age of the earth at 4.5 billion years, our species has existed less than 0.01% as long.

A nice story, but how do we know whether it is true? What evidence supports the theory of natural selection? Let's start with fossils, a term given its current meaning by the German mineralogist Georgius Agricola (1494-1555) in the sixteenth century.[199]

Fossils provide a record of life on Earth.[200] The eponymous event that creates fossils, fossilization, is a somewhat rare event. Many organisms are eaten, leaving no fossil whatsoever. The bodies of others naturally deteriorate from wind and weather. Still more are destroyed by heat, pressure, or other natural processes. Many lack the hard body parts (bones, shells, teeth) most likely to survive fossilization. To be preserved, a fossil must be embedded relatively quickly in a harder, more durable material, such as sandstone or shale. Ash is particularly adept at preserving fossils, protecting them from decay and scavenging. Once a fossil is embedded, there it resides until discovered at some forthcoming date. Earthquakes, landslides, and erosion can expose previously hidden fossils, as can excavation. The circumstances of fossilization, combined with the chance of subsequent discovery, mean the fossil record is incomplete. But, importantly, it is not absent. Let's look at two examples.

In 1946 the Australian geologist Reginauld Claude Sprigg (1919-1994) discovered an accumulation of fossils in the Ediacara Hills in Australia.[201] The fossils, identified as the Ediacara biota, date from 570 million years ago. They represent the earliest known multi-cellular, soft-bodied life forms, mostly with no means of self-locomotion. They thrived for 30-50 million years, then were displaced by arguably the most consequential event in evolutionary history, the Cambrian Explosion.

Though named an explosion, the Cambrian is an interval of time starting 540 million years ago and covering 70-80 million years, long in human terms but fleeting on a geological timescale.[202] The Cambrian marked the first appearance of modern, multicellular organisms and the divergence of most present-day life forms. The extraordinary diversity of the Cambrian includes both soft-body and hard-body (shelly) organisms. An extensive discovery of Cambrian fossils occurred in 1909 when the American paleontologist Charles Doolittle Walcott (1850-1927) discovered a vast trove in the Burgess Shale in Canada.[203]

It is difficult to determine the number of fossils discovered over the last 200 years, but the Ediacaran and Burgess finds provide a clue. There are at least 200,000 fossils from the Burgess Shale available to researchers in various institutions. There is a comparable number from the Ediacara Hills. Similar Cambrian and pre-Cambrian fossils have been found in China, Greenland, and elsewhere. Combined with finds from the post-Cambrian era, the total number of fossils runs well into the millions.

One way to understand the chronology of life is to classify the fossils. This is a task for paleontologists. One classification technique is cladistics, a system of classifying organisms, both living and extinct, based on shared characteristics. Shared characteristics might include the presence of vascular tissues in plants, fused eyelids (versus eyelids that open and close) in lizards, or non-opposable big toes in hominins (a marker of bipedalism).

Traditionally, cladistics was based on the form and structure of organisms, but today extends to molecular characteristics. The idea is that organisms with shared characteristics are more closely related. Paleontologists trace shared characteristics back in time, from ancestor to ancestor, until a point is reached where the characteristic is not present in more distant forebears. This provides a view of the organism's evolutionary ancestry and its relationship to other organisms. Classifying a large number of characteristics from a large number of organisms produces a tree of life, depicting the ancestors and descendants of all organisms. The same process of classification performed by different paleontologists reduces the risk of error but does not eliminate it. And, as you might surmise, to validate the cladistics tree of life it is crucial to know the age of the fossils.

So, when fossils are found, how do we know how old they are? In the preceding chapter, we learned that the age of igneous rocks can be determined by radiometric dating. But fossils akin to those of the Ediacara biota and the Cambrian Explosion often lie in sedimentary rocks. So what do we do? We turn to stratigraphy.

Stratigraphy is a discipline of geology that studies rock layering. Stratigraphists examine various aspects of layering: the direction of layers (horizontal or vertical), the processes that create layers (erosion, deposition), the chemical composition of layers, the types of rocks within the layers, and the orientation of crystals within the rocks. Research performed over the last 200 years reveals that the processes that form the palimpsest of strata and the strata themselves are globally consistent. This allows geologists to determine the sequence in which strata were laid down, establishing the general principle that lower strata are older than higher strata. This principle, combined with an examination of the sedimentary layers

Diagram 7.1
An illustration of rock and fossil layers
determined by stratigraphy.

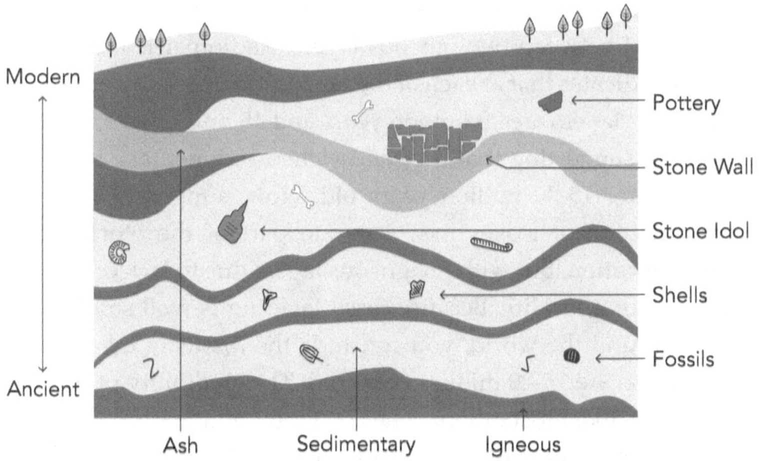

and their embedded fossils, confirms a sensible evolutionary sequence. This principle is depicted in Diagram 7.1.

The order of sedimentary layers provides an indication of relative age (lower strata denote older age), but not absolute age. When a sedimentary layer is sandwiched between igneous layers, however, radiometric dating of the igneous layers produces an estimate of the age of the intermediate sedimentary layer. At a particular geological site, for instance, suppose that radiometric dating shows the lower igneous layer to be 12 million years old, and the upper igneous layer to be 8 million years old. The sandwiched sedimentary layer, and the fossils embedded in that layer, then can be dated to 8-12 million years ago.

A related stratigraphic technique is as follows. Suppose you find ten examples around the world where a sedimentary layer is sandwiched between two igneous layers. Radiometric dating consistently indicates that at each of the ten locations, the lower and upper igneous layers are 20 million years and 15 million years old, respectively, suggesting that the sandwiched sedimentary layer and its fossils are 15-20 million years old. Now, suppose you find a matching zone of fossils in a separate part of the world, at an eleventh location, but with no igneous layers due to lack of volcanic activity at that locality. Because fossil layering is well studied and similar around the world, you conclude the fossils in the eleventh location also are 15-20 million years old. The credibility of this dating method, like most others, relies on large sample sizes.

Another dating method derives from the theory of plate tectonics, a science that emerged in the twentieth century. The British geologist Arthur Holmes (1890-1965) and the German polar researcher Alfred Wegener[204] (1880-1930) originated the theory. Here again is some detail. The lithosphere - the solid mantle and crust of the earth - is divided into several large and small plates. The plates sit on the asthenosphere, a more malleable part of the mantle. The different densities of the lithosphere and the asthenosphere allow the former to slide over the latter, a process called continental drift.[205] Research indicates that the plates, upon which the continents sit, migrate from one to ten centimeters per year. This measure allows the age of landmasses to be estimated. Not their absolute age, but the relative age at which two now-separated continents, for example, once were joined. But wait, there is more.

At some points the asthenosphere is molten. This is the case at mid-ocean ridges, an appellation for underwater mountain chains that encircle the globe beneath the world's oceans. The ridges are formed by magma ascending from the asthenosphere. This process, labeled seafloor spreading, produces new seafloor and new crust as the magma cools. As the magma solidifies, magnetic grains and certain minerals within the rock align to the earth's magnetic field. The alignment, or polarity, is frozen into the rock. Now, an interesting thing happens. The earth's magnetic field flips every few hundred thousand years, north becoming south, and south becoming north. When this happens, new seafloor emerging from the rift records the polarity, which aligns to the new orientation of earth's magnetic field. The study of this phenomenon is termed paleomagnetism. Diagram 7.2 depicts the polarities.

Diagram 7.2
A mid-ocean ridge creating new seafloor
and recording the then-current magnetic polarity.

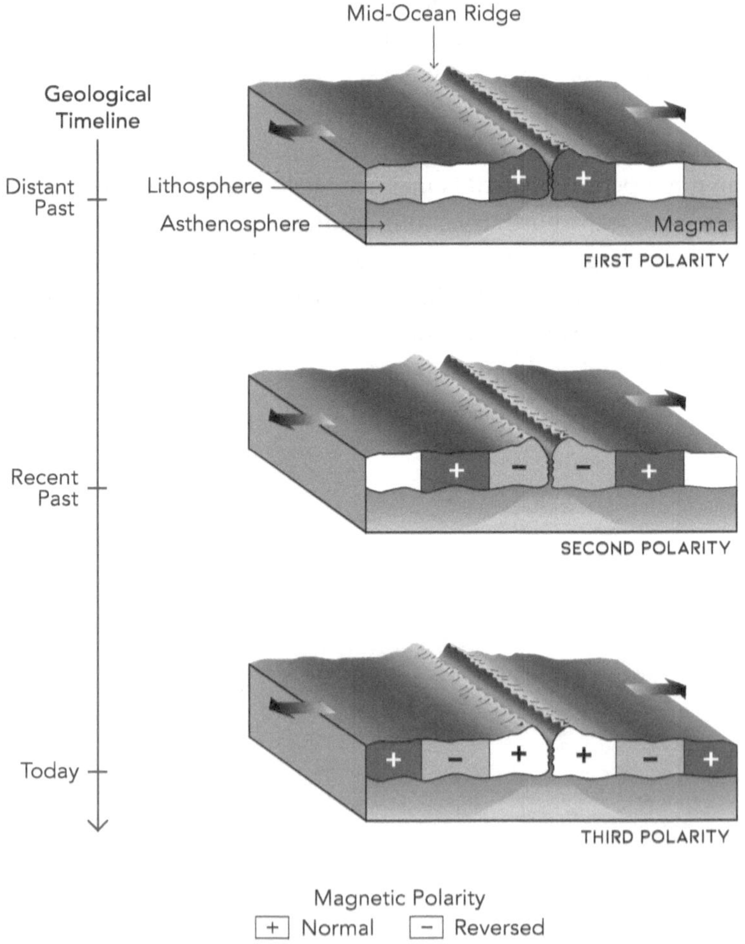

Notice the polarity is mirrored on both sides of the mid-ocean ridge as the sea floor spreads.[206] Now for the denouement. Suppose you find assorted fossils on the east coast of South America and similar-looking fossils on the west coast of Africa. Were they deposited at the same time? The science of plate tectonics attests that South America and Africa once were joined, but began to be pushed apart well over 100 million years ago. The rate of drift provides an indication of age. This fact, along with an examination of matching geological layers, suggests the fossils share a similar chronology. Further, the science of paleomagnetism shows that the igneous rocks on the east coast of South America and west coast of Africa preserve the same mirrored magnetic polarity. This provides independent confirmation that, yes, the fossils on both coasts date from the same period. Diagram 7.3 illustrates this situation.

Plate tectonics credibly measures continental drift back to 200 million years,[207] while paleomagnetism dates seafloor ages back to 160 million years.[208] This is a further case of one clock calibrating or verifying another clock, as discussed in *The Case of Earth's Age*. In this instance, radiometric dating corroborates continental drift, which in turn double-checks paleomagnetism. All serve to date older fossils, while dendrochronology, radiocarbon, and luminescence dating can be applied to more recent fossils.

A routine confirmation of plate tectonics and continental drift occurred in Australia in 2016. GPS (Global Positioning Satellite) map locations are calculated from latitude and longitude, not from the physical position of continents. Because the plate Australia sits on slides north at a rate of seven centimeters per year, the continent's GPS coordinates came to deviate by one meter. To remedy this, Australia officially changed its latitude and longitude coordinates.[209]

Before we leave the dating of fossils, let's take another look at single-celled bacteria, the earliest life forms. One of the easiest to identify is cyanobacteria, which secrete thick cell walls and grow in layered structures. One method of dating cyanobacteria is to observe the structures with an electron microscope, then date the encasing rocks using radiometric dating. Another is to compare the

Diagram 7.3
An illustration of mirrored magnetic polarity
and matching fossil records as determined by stratigraphy
and paleomagnetism.

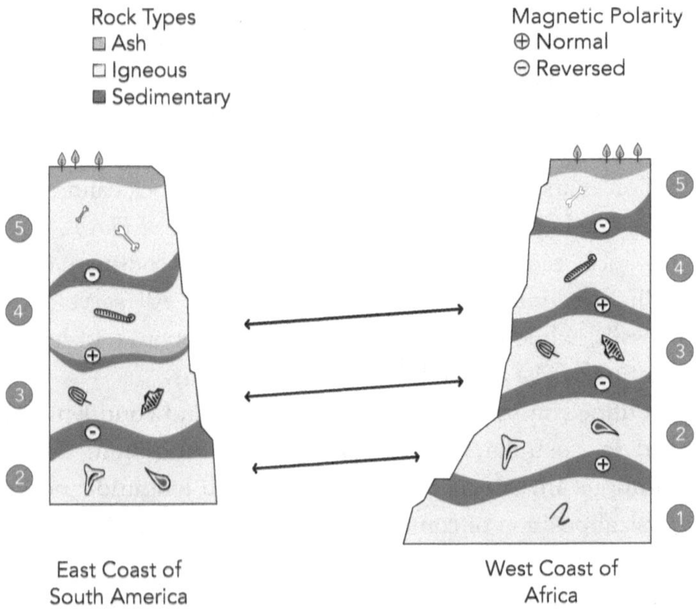

levels of Carbon-13 to Carbon-12 in a sample (carbon from non-living sources contains a higher ratio of Carbon-13 to Carbon-12 than does carbon from organisms capable of photosynthesis). These techniques indicate that the earliest cyanobacteria arose about 3.5 billion years ago. This yields another data point on the age of the earth, as the oldest living organisms are unlikely to be older than earth itself.

In this and the previous chapter, I presented a variety of dating techniques. I will resist the urge to describe still more. Suffice it to say many others exist. Each serves to date objects, confirm sound hypotheses, or refute hollow claims. The following gives a sense of their variety and the age (in years) to which they yield credible results: ice layering (150,000); y-chromosome mutation (350,000);[210] erosion (6 million); seabed plankton layering (50 million); impact crater frequency (300 million);[211] slowing of the earth's rotational speed (600 million); the recession rate of the moon (4 billion);[212] and gravitational lensing (14 billion).[213] Still other measures of gradual natural processes - permafrost formation,[214] wood petrifaction, rock varnish accumulation,[215] and calcium carbonate stalactite growth[216] - not only disprove all young-Earth hypotheses but, in like manner, corroborate the timescales necessary for an evolutionary theory of life.

Fossil identification and classification, stratigraphy, paleomagnetism, and continental drift all support the theory of evolution by natural selection. All offer testable predictions, and none has been disproved. But what might disproof look like? My favorite example comes from Dawkins. Our story of life indicated that mammals arose some 200 million years ago. Suppose, however, that someone chanced upon a mammalian fossil from the Devonian, a geological period 200 million years older still. Such a find, of a mammalian fossil older than the existence of the first mammals, would irrefutably indicate that the existing premise of mammalian emergence is wrong. Period. No if's, and's, or but's. Just wrong. Yet not one mammal fossil has ever been found in Devonian rock or any older strata. They are not just statistically rarer in Devonian rock; they do not exist at all. Not one.[217]

Shifting our attention from fossils to living organisms, the mountain of evidence in support of natural selection grows higher still. Study after study corroborates the adaptation of organisms over multiple generations in response to their environment. Instances include the diversification of lizards on Croatian islands;[218] the colorization of guppies in mountain streams in Trinidad, Tobago, and Venezuela;[219] the development of alcohol tolerance in fruit flies;[220] the evolution of wings, tails, bills, and legs in non-native bird species in Hawaii;[221] and the domestication of wild foxes in Russia[222] - to cite but five of many, many examples. Recall, too, that in Reason, scientists publish their methods and results. Studies, including the above, are transparent. Anyone willing to expend the time and effort (and money) can repeat the experiments to verify the results.

One of the more compelling studies began at the University of Michigan in 1988. A team chose *Escherichia coli*, a bacterium that reproduces asexually and is easy to clone, to initiate a long-term experiment on natural selection.[223] Diminutive organisms, from their rates of reproduction and speeds of adaptation, offer microbiologists fascinating areas of study. The team inserted *E. coli* into twelve test tubes, each containing a nutrient broth with glucose as a food source. Each of the twelve populations first catapulted, then declined, as the food source was exhausted. Each day, a sample from each tube was placed in a corresponding new tube with fresh nutrients. The team has continued this experiment for 30 years, creating over 67,000 generations of bacteria to date. During this time, the team observed several notable changes. In one instance, bacteria better able to make use of glucose increased in body size and were able to reproduce faster, dominating their population. In another, bacteria developed the capacity to use citrate as a new food source, an ability traced back to changes at the molecular level. The experiment confirmed essential elements of evolution: random change followed by non-random adaptation to the environment.

Even stronger evidence for natural selection comes from a source not well understood in Darwin's time: genetics. Genetics is the study of genes and genetic variation. The father of genetics is the friar Gregor Johann Mendel (1822-1884), who was born in a corner of the Austrian Empire that today is the Czech Republic. By conducting experiments with pea plants over 8-10 years, Mendel was able to establish the basic rules of heredity.[224] The birth of modern genetics, however, occurred in 1953 when British biologist Francis Crick (1916-2004) and American biologist James Watson (1928-) determined the structure of DNA.

It now is understood that DNA consists of nucleotides, four chemical molecules arranged into long sequences. Think of DNA as a code and the four nucleotides as letters of the code. The number and arrangement of the letters direct (via several intermediate steps) the growth and traits of an organism. A gene is a segment of DNA, a unique configuration of nucleotide letters that forms the molecular unit of heredity. The number of genes, the number of letters within each gene, and the ordering of the letters vary by organism, but the code itself is the same: from bacteria to bears, from fungi to flowers, from jellyfish to giraffes, and from hippopotami to humans. Genes contain the sets of instructions that determine the development, functioning, and reproduction of all living organisms. The number of genes in the human genome, for example, is approximately 20,000, each gene containing from several hundred to several million pairs of letters.

Genetic paleontology explains evolution and natural selection in enormous detail. In cladistics, genetics more accurately explains the relationships between living and extinct organisms. In heredity, it explains how information is transmitted from generation to generation. It deciphers mutations as the re-configuration of nucleotides, a regularly-occurring, dynamic process. It elucidates the evolutionary relatedness of every major group of species to every other from their genetic composition. It confirms all organisms from bacteria to *H. sapiens* use the same DNA code, each species expressed by what is written in the code. And it attests that all life forms operate from a common genetic model derived from a common ancestor.

Even if no fossils had ever been found, genetics thus provides the most robust evidence of both evolution and natural selection.[225] Perhaps of even greater interest, evolutionary genetics explains how genes act to spread themselves in our species: by constructing our brains so that we enjoy life, health, relationships, and sex. Genes thus increase their own likelihood of spreading to the next generation, transporting with them our inherited characteristics.

This is a good point to comment further on *H. sapiens*. We are hominids, part of the family of gorillas, orangutans, bonobos, and chimpanzees. We are a close family, so close that we share roughly 98% of our DNA with our chimpanzee cousins - as measured not by chromosomes or genes, but by the number of matching DNA letters (base pairs).[226] We are one of an estimated 15-20 bipedal species to have emerged in the last two to six million years.[227] Other bipedal species include *Australopithecus afarensis* (southern ape from Afar), *Australopithecus africanus* (southern ape from Africa), *Homo habilis* (handy man), *Homo erectus* (upright man) and *Homo neanderthalensis*[228] (man from the Neander valley in Germany). Bipedalism, by the way, refers to *habitual* bipedalism - species who walk on two legs as their primary mode of travel.

Genetics and the fossil record provide a means to differentiate among bipedal species. Earlier bipedal hominins are characterized by smaller brain sizes, protruding facial features, and a larger dentition. Fossils also provide evidence of diet (via the teeth) and bipedalism (via limb bones). But what led to bipedalism is not yet clear. Perhaps natural selection favored the more efficient cooling an upright posture provides. Perhaps it was the facility to run faster or see farther. Or perhaps climate events altered habitats and food resources which, in turn, prompted evolutionary change. What led to our survival and the demise of other hominins, here again, is unclear. But our emergence 200,000 years ago, while relatively recent, allows me to write this book and you to read it.

In hindsight, it may seem obvious that natural selection is the mechanism that explains evolution. But Darwin and Wallace were not the first or last to offer a conjecture. Reason and Religion both have put forward competing speculations. From Reason, the French naturalist and soldier Jean-Baptiste Lamarck (1744-1829) suggested that useful traits are inherited - when giraffes stretch their necks to reach leaves, their offspring inherit longer necks.[229] Later, in 1844, the Scottish geologist Robert Chambers (1802-1871) published *Vestiges of the Natural History of Creation*, which argued for a cosmic source of evolution. The work first was reviled by clergy for contradicting natural theology, then by scientists for its speculations and want of evidence.

Other ideas from Reason, past and present, include preformationism, which contends that both egg and sperm contain tiny, miniature babies;[230] mutationism, which submits that evolution proceeds by the sudden emergence of new life forms; and orthogenesis, which asserts that evolution is driven by forces internal to the organism and not by environmental pressures.[231] Mutationism and orthogenesis, while offering possible explanations for infrequent, one-off events, do not offer a broad-gauged explanation of evolution. The evaluation and refutation of such speculations once again evince the ability of the scientific method to separate the wheat of viable theories from the chaff of fanciful ideas.

From Religion, creationists contend that a deity designed each species for its purpose, remaining unchanged to this day. To defend this position, the English clergyman William Paley (1743-1805) offered the popular watchmaker argument, which can be briefly stated as follows.[232] If one finds a watch, having never seen one before, its complexity would suggest a *sui generis* intelligent designer, just as the complexity of life and the universe itself suggest an intelligent creator. This argument, without closer scrutiny, offers superficial appeal. Yet it has been disproved multiple times, the most obvious refutation deriving from natural selection. The flaw in the argument, of course, is that watches cannot reproduce themselves. If a watch could replicate itself to a close but not perfectly accurate degree, it would, over time, lead to the creation of other complex objects without input from or necessity for an intelligent designer.

ACCEPT

Among all the areas where Religion disagrees with Reason, the discord perhaps is greatest with evolution. Religion struggles with the theory of evolution by natural selection, as it contradicts a core teaching: that one or more deities created life in its myriad forms, each attuned to its respective environment, each perfect in its design, each unchanging from generation to generation. Natural selection thus refutes the creation myth of nearly every faith-based belief system. Undaunted, Religion employs its *how to think* process to proclaim, declaim, and acclaim the validity of its creation myths. Yet, once again, it offers few testable predictions, adduces no empirical evidence, and tenders no independently verifiable claims. Its followers are simply asked and expected to accept the truth of its pronouncements by virtue of appeal to authority.

Evolution by natural selection, via the contrast it exposes between Religion and Reason, provides a number of insights on the fifth step of the scientific method, Accept. As we have seen, in Reason a valid theory will have passed a range of testable predictions. It will have explained all the observations of reality that preceded it. And it will have correctly predicted the results of new observations. The weight of such evidence results in acceptance of the theory. Rather than acknowledge this reasonable approach, Religion creatively employs numerous methods of non-acceptance. Here follow several.

ELEVATE FAITH ABOVE REASON

The first method of non-acceptance relies on sacred texts and sacred persons to elevate faith above Reason. The philosopher and theologian, Augustine of Hippo (354-430), whose copious writings influenced the progression of Christianity, set the tone of fideism. He warned against astronomers, mathematicians, and others who ventured to predict the future, inferring they might be in league with demons.[233] He believed the future was the sole province of his deity. Similarly, the Benedictine monk and cardinal Pier Damiani (c.1007-c.1072) saw the rise of universities in his era and their fledging methods of teaching as a threat to well-established, faith-based medieval beliefs.[234]

A further assault on Reason issued from Pope Pius IX (1846-1878). In 1864 Pius published his *Syllabus of Errors,* condemning eighty errors of thought that contravened Church doctrine.[235] Error number eight on his list is ranking Reason equal to Religion. To bolster the authority of both Pius IX and the Church, in 1870 the First Vatican Council declared the pope infallible. Anyone who challenged the word of the Roman Pontiff was, and is, deemed "anathema" - cursed.[236] More recently, in 2009 Pope Benedict XVI (1927-) reaffirmed, "Reason always stands in need of being purified by faith."[237]

The *Catechism of the Catholic Church* provides another contemporary example: "Though faith is above reason, there can never be any real discrepancy between faith and reason. Since the same God who reveals mysteries and infuses faith has bestowed the light of reason on the human mind, God cannot deny himself, nor can truth ever contradict truth. Consequently, methodical research in all branches of knowledge, provided it is carried out in a truly scientific manner and does not override moral laws, can never conflict with the faith."[238] Within the quote, note the words *provided it ... does not override moral laws.* Conveniently and unsurprisingly, Religion positions itself as the arbiter for adjudicating whether secular inquiries override moral laws, exercising its jurisdiction wherever and whenever it deems appropriate. A present-day example concerns bioethics, where Religion attempts to exert its moral certainty on cloning, gene therapy, surrogacy, birth control, gay marriage, and transgender rights.

Christianity possesses no monopoly in ranking faith above Reason. Other religions duplicate the practice, elevating faith to the apex of their theologies. The maneuver invariably relegates Reason into subservience. Sikhism, for instance, states that God is "beyond comprehension, even if contemplated hundreds of thousands of times."[239] Hinduism, in turn, refers to Brahma, the ultimate reality, as "Beyond all thought ... argument and illustration, unknowable and causeless."[240] And Islām declares that, "Whoever seeks knowledge for a reason other than the sake of Allāh, or intends it for a purpose other than for the sake of Allāh, let him take his place in hell."[241]

By exalting faith, Religion accords itself the right to reject any noisome findings from Reason. Theologians extend the charade by depicting odious evidence as trials of faith. In this scenario, the disciple acknowledges the existence of loathsome evidence: that natural selection acts as a choosing agent; that fossil records describe trait inheritance; and that radiometric dating denotes a multi-billion year age for the earth. But the apostle interprets the evidence as a ruse, created by a deity to assail and entrap the apathetic or perfidious mortal. Thus armed, the tractable zealot not only rejects the evidence, but glorifies the act of repudiation as a steadfast declaration of faith. This technique allows the proponent to disregard or derogate any facts, people, or practices that raise uncomfortable truths. Scriptural interpretations often reinforce the technique, as underscored by the Biblical passage, "For whatsoever is not of faith is sin."[242] Yet, particularly for belief systems that embrace pre-destination, it is unclear why an impish or devious deity would set up trials when the outcome is foreordained and inevitable.

Another scenario leverages the locution of Religion to arbitrarily terminate inquiry. Rather than evaluating or critiquing distasteful evidence, Religion viscerally responds with rehearsed catchphrases to sidestep the evidence: the Lord moves in mysterious ways; man is unable to comprehend the mind of God; it is the will of Allāh. The adages, aphoristic to a believer yet specious to an outsider, discourage further consideration and analysis. All such adages convey a shared connotation to the secularist: We don't know, we don't want to know, and you should not want to know.

ELEVATE INTUITION ABOVE REASON

As with faith, Religion also attempts to elevate intuition above Reason as a source of truth. We outlined the concept of intuition in *The Origin of Religion*. To quickly recap, we:

- Referred to intuition as a path to spiritual knowledge.
- Noted that intuition is intimately intertwined with inspiration, emotion, and feelings.
- Cited scriptural interpretations affirming intuition as a source of knowledge.

If two of the five sacreds - deities and scripture - assert the legitimacy of intuition as a source of truth, the remaining three propel its application. Witness instances within modern-day revivalist Evangelicalism, where a sacred person (a minister) conducts a sacred ritual (a sermon) in a sacred place (a megachurch). The minister passionately preaches the word of the Lord, liberally interspersing his exaltations and castigations with quotations of Biblical verse. He recites well-rehearsed shibboleths: God is love; feel the spirit of the Lord; allow His almighty presence into your heart; trust in His will. He praises the Lord's benevolence, lauds His compassion, and extols His forgiveness. Music and lighting effects, orchestrated to heighten emotions, accompany the spiritual celebration. An audience of hundreds or thousands, eyes closed, heads upturned, arms raised, sway left and right in unison, absorbing the professed truth. Remote congregants imbibe the awakening, streamed live to computer screens, tablets, and mobile phones. The predictable outcome? A multitude of enraptured, spiritually-reborn disciples, their beliefs authenticated, their devotion reinvigorated, their souls cleansed - all swayed by intuition, emotion, and passion to embrace the truths of the Lord.

Revivalist Evangelicalism represents but one of many attempts to rationalize intuition and emotion as sources of truth. Consider another instance where communal worship is structured not only to strengthen doctrinal commitment but to encourage feelings-as-truth. In Mormonism, members periodically bear testimony at worship services, often on the first Sunday of each month. An individual stands among the seated congregation, head downcast, and emotionally declares their love of Christ and their belief in the teachings of the Church, often with breaking voice and tearful eyes. No one is required to bear testimony, but the willingness of one to perform the act inspires others. The congregation interprets the perfervid declaration of belief as a testament to the truth of the faith.

But do intuition, emotion and feel-goodism convey truth? Here, social psychology, the study of individual behavior in social contexts, offers some insights. One is "motivated thinking," a phenomenon where conscious and subconscious mental processes interpret information in a manner that aligns with one's preferred values, attitudes, or outcomes.[243] Research suggests that motivating thinking affects how, and how much, we reason. For example, we quickly accept favorable evidence but apply greater scrutiny to unwelcome information. Among other findings, motivating thinking affects how we organize knowledge in memory, how we recollect it, and how we attribute cause and effect. Taken together, these traits often instill bias, serving to justify one's beliefs and decisions. Motivated thinking thus tends to elevate emotion over facts, a characteristic that dovetails flawlessly into Religion's evidence-free, intuition-as-truth assertions.

Reason, in contrast, recognizes that intuition and commonsense notions are simply ill-equipped to disclose the nature and detailed-working of our world. While Reason's practitioners might begin their inquiries from an intuition, they never treat intuitions as proof. Yet Religion cannot accept this answer. It resists efforts to replace vague intuition with unambiguous precision. It refuses to dissect both intuition and the "truths" obtained from intuition into component parts, each to be assessed, each to be confirmed or refuted, solely on the basis of objective inquiry. As such, it must rely on non-reasoning, non-intellectual, and non-empirical stratagems not only to dispel uncomfortable and bothersome facts, but to assert its own truths, each unencumbered by evidence. And it must buttress its position with pronouncements from ancient pundits, such as Augustine of Hippo (354-430), who declared that divine truth is inscribed in the human mind.[244] In this manner, Religion seeks to install intuition, emotion, and feelings above Reason as the arbiters of truth and reality.

ESTABLISH SEPARATE REALMS

The *God of the Gaps* argument forms a third method of non-acceptance. The argument claims that any gaps in knowledge unexplained by Reason *ipso facto* attest to the existence of a deity. This, in essence, divides knowledge into separate realms of Reason and Religion. By proclaiming the gaps as the realm of a deity, Religion maintains its authority on metaphysical and theological subjects.

It certainly is true that Reason cannot yet explain all phenomena. Ideas and hypotheses might exist with regard to specific enigmas, but in many instances there are no accepted theories. Reason, for example, cannot explain the origin of the Cambrian Explosion. Perhaps a climatic event, driven by glaciation or volcanism, accelerated evolution. Perhaps changes in oxygen levels in the atmosphere drove genetic modification. Perhaps internal genetic changes, such as the ability to better process food for energy, boosted the diversity of organisms. Perhaps plate tectonics altered sea levels and sea chemistry, creating more attractive habitats. Or perhaps it was a combination of all these ideas, or something else entirely.

Similarly, what specific factors caused *H. sapiens* to arise? What factors enabled *H. sapiens* to survive, while the other 15-20 species of our bipedal, hominin cousins went extinct? Perhaps our species was better able to adapt to changing habitats. Perhaps our larger brain size enabled us to out-compete other species. Again, Reason has hypotheses, but no definitive answers. This permits Religion to step in with its own fables, each fashioned from a sacred text, each attributed to its deity, and each asserted as eternal truth. This is not difficult, of course, as Religion simply invents its tales. And should its fabrications be challenged, Religion responds with a shopworn expedient: It simply appeals to authority.

The *God of the Gaps* argument is an example from a broader category of fallacious reasoning known as *argument from ignorance,* which claims a proposition is true if it has not been proved false. The argument is specious on multiple grounds, but we will focus on three. First, the argument is a point-in-time assertion. It overlooks the progress of Reason. On this point, consider a medieval monastery in 1400 CE that possesses a fragment of wood. The abbot claims the relic is part of the true cross of Christ. At that moment in time, there was no reliable method for doubters to disprove the claim. Now there is. If dendrochronology confirms the wood came from a tree that grew in 1200 CE, the claim is refuted. Religion fails to acknowledge that ample evidence might not yet exist to confirm or disprove a claim, but soon could.

Second, the argument relies on an imbalance of proof. Here, recall that Religion demands absolute and irrefutable evidence from Reason, but holds itself to no similarly robust benchmark. It undertakes no effort to objectively and rigorously produce positive evidence of its own claims. It merely asserts its claims by fiat. And by deliberately structuring its claims to avert testability, it lessens the risk of refutation.

Third, and perhaps surprisingly, the *God of the Gaps* ploy to enjoin a foothold for Reason has come to haunt Religion. The dilemma is simple. Envisage a continuum of all possible knowledge. One point on the continuum represents, say, the knowledge of comets - their origin, their composition, and their movements. Throughout most of human history, the presence of a comet in the night sky marked an unexplained phenomenon, often interpreted as an omen from a tri-omni being. Today, however, astronomy and its sub-discipline, astrophysics, can explain the genesis and motions of comets, obviating both mystical and theistic interventions. The purported deity plays no role. As Reason advances, this pattern repeats. Each new discovery squeezes the illusory assemblage of gods and goddesses into an ever tinier gap on the continuum. For Religion, the end state is unattractive.

DISMISS EVIDENCE

A fourth method of non-acceptance is simply to dismiss evidence, yet offering none in return. The story of Noah's Ark and the mythical flood typify this method. Biblical creationists believe God saved one family and all the world's animals by instructing the individuals to build a sizable vessel, an ark. The family collected examples of each species, herded them into the ark, and survived a lengthy flood, while the rest of humanity, and all the earthly fauna and flora,

Diagram 7.4

The *God of the Gaps* argument, where
gaps in knowledge unexplained by Reason are deemed
proof of the existence of a deity.

Continuum of all Possible Knowledge

■ Explained by Reason

■ Unexplained by Reason, deemed by Religion as proof of a deity

were destroyed. As the flood waters receded, the ark settled in the Mountains of Ararat, thought to be in present-day Turkey. Its occupants disembarked to repopulate the world.

What evidence might religious apologists provide to validate this story? Consider the following. Once the ark settled on dry land (I borrow heavily from Dawkins here[245]), why would all marsupials - pouched mice, koalas and kangaroos - make their way from Ararat to Australia, yet to nowhere else in the world? What path did they follow across land and sea? Why are there no fossil records of their journey - no bones, no footprints, no hides, no petrified feces? Why are there no traces whatsoever of their divinely-inspired odyssey? Why did some members of the migration not stop to settle in India or Indonesia? Why did no placentals, animals which nourish their fetuses from a placenta, join the migration to Australia? Why did all species of armadillo journey to South America and to nowhere else? Why did all penguins waddle and swim to Antartica, not one traveling to the equally chilly Arctic? How did the millions of species of bacteria wiggle their way back to their respective ecological niches around the world? For that matter, how did Noah collect them in the first place? Why is the narrative silent on flora? How did all the species of trees, bushes, grasses, and flowers survive? What befell the fresh-water fishes, who would not have survived in a globally salty sea? How were the thousands of species of fresh-water marine life, often unique to specific lakes and rivers, collected from around the globe then returned to their singular habitats when the flood subsided? The tale of a global flood offers opportunities without end to draft hypotheses, collect empirical data, and provide evidence both of its occurrence and its consequences. Yet Religion offers none. No objective evidence of any kind. The explanation is simple. It never happened.

ALTER EVIDENCE

When it becomes impractical simply to dismiss unwelcome evidence, a fifth method is to alter its meaning, so that it conforms to religious doctrine and ratifies spiritual belief. An example of this form of natural theology is the *Omphalos: An Attempt to Untie the Ge-*

ological Knot, written by the English naturalist Philip Henry Gosse (1810-1888). In *Omphalos*, Gosse extends the argument of Adam's navel, which can be described as follows. If Adam were born naturally, he would have a navel. If a deity created Adam, he would not need a navel and might not have one. So did he have a navel or not? (Yes, theologians actually debate this.) Gosse argued that Adam had a navel, and that the navel was *prochronic* - occurring outside of conventional time. Gosse further argued that when God fashions a creature, the creature cannot help but display the semblance of preexistence, thus completing the circle of time from prochronic to *diachronic* - existence in conventional time. He then cited several hundred examples of prochronic tendencies. If God created a boar one hour ago, for example, it would exhibit worn molars showing years of use.

Gosse did not constrain himself to navels and molars, but further extended his argument to the entire earth and its fossil record. He claimed the fossil record is a prochronic event, placed through the agency of a deity to portray the façade of previous life, and to be consistent with fossils created in the diachronic era.[246] I award Gosse high marks for imagination and creativity, and for furnishing another superb instance of natural theology.

CLEANSE EVIDENCE

A sixth method is to cleanse evidence, selecting kernels that uphold dogmatic tenets while ignoring incompatible data. Two facets of nature illustrate this method: its complexity and its beauty. Religion regularly attributes the intricacy and elegance found in nature as evidence of creation by a deity, yet routinely dismisses or ignores conflicting evidence. We already have seen how natural selection can produce organisms of incredible complexity, without invocation of a deity. The argument for beauty differs.

It is certainly correct that nature offers fantastically beautiful phenomena, but it also offers incredibly harsh and grotesque examples. When asked by creationists why he does not credit a deity for the beauty in the world, the naturalist and broadcaster Sir David Attenborough (1926-) responded, "They [creationists] always mean beautiful things like hummingbirds. I always reply by saying that I think of a little child in east Africa with a worm burrowing through his eyeball. The worm cannot live in any other way, except by burrowing through eyeballs."[247] Eyeball-burrowing worms (caused by the larvae and adults of the nematode worm *Loa loa*), skin-burrowing botfly maggots, famines, volcanic eruptions, earthquakes, and tsunamis are but a few counter-examples to the beauty in the world. But Religion rarely "credits" such events to its deities (apart from cases of punishment). Instead, its apologia ascribes only propitious acts to its gods and to its beliefs.

DISCREDIT EVIDENCE

A seventh method of non-acceptance attempts to discredit Reason via false analogy. A quote from an early candidate in the 2016 Presidential election in the United States provides a superb example: "What you're telling me is, if I blow a hurricane through a junkyard enough times, over billions and billions of years, eventually, after one of those hurricanes, there will be a 747 fully loaded and ready to fly."[248]

The analogy, which endeavors to deride the theory of natural selection, offers three instructive insights. First and foremost, it postulates a "straw man" analogy, a false analogy constructed on superficial and misleading comparisons. In this instance, the jetliner analogy fails to capture key elements of natural selection: that organisms (or jetliners) can reproduce themselves with variations from generation to generation; that organisms evolve slowly over prolonged periods of time rather than arising fully-formed from a single windstorm; and that fossil and genetic records document a chronology of change. Second, the advocate ascribes the false analogy to Reason, when in fact Reason makes no such fatuous claims of self-assembling airplanes. And third, the advocate then ridicules and disproves the spurious analogy, positioning the disproof as ratification of religious belief. Unfortunately, this three-step recipe - inventing a false analogy, ascribing the false analogy to the opposition, then disproving the false analogy - is a common and effective technique, whether practiced through nescience or as disinformation.

AMEND ETERNAL TRUTHS

When all else fails, when it becomes too foolish to discredit, dismiss or alter overwhelming evidence, Religion turns to an eighth method of non-acceptance: amendment of its eternal truths. As previously noted, it can be difficult for Religion to claim immutable truth, then alter it. Yet Religion has walked this path numerous times, most notably when its preternatural claims are shown to be false. We already cited examples of such revisions: Most religions now accept that the earth travels around the sun and that planets orbit in ellipses. Other faiths, unable to repudiate the Cambrian Explosion, now cite it as evidence of God's creation. And more than one religious scholar now accepts the theory of evolution by natural selection, but claims that a deity designed and initiated the mechanism. This stratagem maintains the godhead in the role of original designer, with nature operating to the dictates of the design.

In every instance, Religion's method of amending its eternal truth is simple. A sacred person simply proclaims a new revelation, ostensibly without guile or embarrassment, and distances his religion from the previous belief.

The practitioners of natural theology employ the eight methods cited above, plus others, to repudiate unwanted evidence. The outcome, all too often successful, is an amalgam of misinformation and disinformation designed to preserve the relevancy of the faith and the convictions of the faithful. Zealots press further, wittingly extolling a positive correlation between distasteful evidence and resolute faith.

I choose to close this section with an anecdote from Pascal Boyer, an American anthropologist and evolutionary psychologist, that typifies the myriad and creative techniques employed by Religion to circumvent acceptance of unwelcome evidence. At a dinner in a Cambridge college, Boyer related some of the beliefs he has encountered over the course of his anthropological studies, among them a belief in witches who feast on their victims at enormous banquets, supported by eyewitness claims of the witches flying overhead in the night. A guest at the dinner expressed interest in the stories, yet wondered "how people can believe such nonsense." The guest was a Catholic theologian.[249]

A recap of the eight techniques outlined in this section appears in Diagram 7.5

In the comparison of Religion and Reason, it is important to remind ourselves that Reason does not produce a catalog of absolute knowledge. This includes natural selection, various elements of which Reason continues to vigorously debate. Here is a sampling.

Does natural selection operate solely at the level of the gene or sometimes at the level of the organism? Does natural selection proceed only at a gradual, constant rate, or does it sometimes proceed in spurts and jumps (the idea of punctuated equilibrium, which asserts that notable evolutionary change can occur in selected circumstances, but the fossil record is more likely to record stable periods)? What is the relative importance of vertical versus horizontal gene transfer? To what extent does female choice, versus competing causes, explain trait inheritance? In our species, what is the relationship between the physical structure of the brain and the less tangible essence of the mind? What is the connection between the origin of language and the evolution of consciousness? When and why did consciousness arise? To what degree are memes (cultural units of heredity) subject to natural selection? Do memes explain behavioral aspects, such as our attachment to cultural traditions? Will the discovery of life on other planets reveal an alternative mechanism to DNA and suggest a dramatically different tree of life? These and other questions are debated, sometimes hotly, in search for answers.

Diagram 7.5
Examples of methods of non-acceptance employed by Religion, through natural theology and appeals to authority, to repudiate unwelcome scientific evidence.

Natural Theology	Appeals To Authority
Alter Evidence	Elevate Faith Above Reason
Cleanse Evidence	Elevate Intuition Above Reason
Dismiss Evidence	Establish Separate Realms
Discredit Evidence	Amend Eternal Truths

But the goal is to improve the catalog of knowledge. There is no suggestion that answers to these questions will willy-nilly overturn more than 100 years of evidence supporting the theory of evolution by natural selection.

Most of the examples of Reason in this book demonstrate inductive reasoning, but the theory of natural selection also provides an excellent example of abductive reasoning, that is, disproof by alternatives. There simply is no other theory, no other explanation, that accounts for the broad range of evidence from geological paleontology (fossil records, taxonomic relationships, stratigraphy, plate tectonics) or from genetic paleontology (mitochondrial DNA, nuclear DNA, gene sequencing, gene amplification).[250] This also demonstrates the principle of Ockham's Razor, articulated by the English friar and philosopher, William of Ockham (1287-1347). Ockham asserted that, among competing hypotheses, the one with the fewest assumptions that best fits observations should be selected.[251] Inheritance of adaptively-advantageous traits - a beautiful instance of the principle. Ockham, incidentally, penned several treatises challenging the authority of the pope. As you by now will have surmised, the Church retaliated with a well-rehearsed tactic: suppression by excommunication.

In *The Case of Gravity*, we attributed humanity's first great displacement to Copernicus, who dislodged our species from the center of the universe. Darwin, 316 years later, was the author of the second, ousting humanity from its privileged position as an intentionally designed, essential species. Natural selection affirms we are the product of an unintelligent process, a complex and iterative response to our environment through a succession of adaptations, with no supernatural design or divine guidance. It is natural selection, resulting in species that are well enough but not perfectly adapted to their habitats, which produces the boundless complexity and diversity of life. Yet here we are, *H. sapiens*, the current culmination of an incredible journey that began 4.5 billion years ago on our home planet. It's an amazing story.

We close this chapter by presenting the conclusions from *The Case of Human Origins* for the final step of the scientific method, Accept, in Diagram 7.7.

MISCELLANY

The idea of splitting intuition into its component parts comes from the Swiss psychiatrist Carl Jung (1875-1961), the founder of analytical psychology. Jung stated, "Intuition is an irrational function, notwithstanding the fact that many intuitions may subsequently be split up into their component elements, whereby their origin and appearance can also be made to harmonize with the laws of reason."[252]

Neanderthals prospered between 400,000 to 40,000 years ago as a separate species from modern humans. Genetic evidence confirms that *H. sapiens* interbred with Neanderthals before the latter species became extinct.[253]

In addition to Crick and Watson, many others contributed to the discovery and understanding of DNA, among them the Russian physicist George Gamow (1904-1968), the English chemist Rosalind Franklin (1920-1958), and the British physicist Maurice Wilkins (1916-2004).

This chapter simplifies the story of genetics by ignoring recent advances in gene sequencing, epigenetics (changes in gene expression without altering DNA), and evolutionary development (the integration of evolutionary biology with developmental genetics). Together, these fields are unveiling the complex interplay between genes, environment, and evolution.

Vertical gene transfer (VGT) refers to the transmission of DNA from parent to offspring. Horizontal gene transfer (HGT), in contrast, refers to transmission from non-parental sources. In recent decades, researchers have identified several types of HGT. One is the uptake of naked DNA, where a cell absorbs free-floating genetic material from the rupture of a nearby cell. Another is viral transfer, where a virus picks up DNA in one cell, then deposits it in another. HGT can occur both within and across species boundaries. From this distinction, note three points. First, consider the convenient "tree of life" metaphor, where dissimilar groups of flora and fauna populate each limb of the tree. Cladistics and genetics inform the ancestor and descendant relationships among the limbs. VGT and HGT, in combination, suggest a more tangled metaphor. Life is less a tree and more a collection of vines, each twisting upwards, downwards, and sideways. Second, the vine metaphor suggests life arose not from a single - or a small number of - ancient and distinct microorganisms, but from a messier conglomeration of primitive cells. And third, with regard to natural selection, HGT represents a source of genetic variation that is subject to the same environmental pressures as VGT. See Diagram 7.6.

Diagram 7.6
Metaphors for the evolution of life.

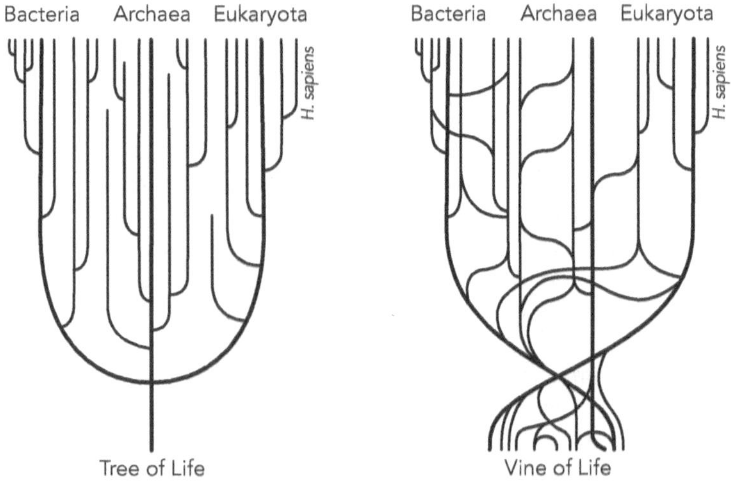

A summary of the different approaches of Religion and Reason in the final step of the scientific method, Accept.

Diagram 7.7

Principles

	Religion	Reason
① ASK	Compels adherence to dogma; suppresses, even punishes, inquiry that challenges existing beliefs and convictions.	Encourages a culture of openness, advocates inquiry that challenges established knowledge.
② PREDICT	Deliberately structures beliefs to avoid testable predictions, posits faith as superior to evidence.	Demands hypotheses with testable predictions that can be confirmed or disproved by observation and experiment.
③ OBSERVE	Selects observations that ratify its beliefs; dismisses, disparages, or condemns contradictory evidence.	Conducts observations and experiments, analyses results to accept or reject hypotheses.
④ VERIFY	Relies on appeals to authority and circular reasoning, professes scripture as truth then argues truth from scripture.	Requires independent verification and a preponderance of evidence; dismisses the authority, prestige, and intelligence of the advocate.
⑤ ACCEPT	Refuses to accept conclusions that contradict its beliefs, again invoking appeals to authority to preserve its ideology.	Accepts verified theories into its compendium of knowledge, but remains open to future challenge and revision.

[VIII]

Belief

A man with a conviction is a hard man to change.[254]

LEON FESTINGER (1919–1989) AMERICAN PSYCHOLOGIST

The cases on *Gravity, Earth's Age,* and *Human Origins* revealed various facets of Religion's *how to think* toolkit, one that relies on revelation, scripture, theistic authority, and related elements from the five sacreds as sources of truth. Yet the toolkit, as demonstrated, suffers from innumerable defects. It forces Religion, ever eager to shield its ideology from critical scrutiny, to obstruct inquiry. It generates no empirical evidence to substantiate its claims. It shuns testable predictions. In unwonted cases where it offers predictions, its prophecies are routinely refuted. It offers no independent verification. It refuses to accept outcomes that contradict its teachings. It plies myriad ruses to dismiss unpalatable evidence. It obliges Religion to structure its dogma to evade testability. It shelters behind the notion of faith to safeguard ingrained convictions. And it induces each denomination to assert that its deity, and its deity alone, conveys truth, while at the same time dismissing the thousands of other purported deities.

This litany of faults, ruinous to any other discipline but endemic to Religion, begs an elemental question. Why do people persist in believing tales of divine creation and supreme beings? What attracts humans to legends that arose in the Bronze Age, an era of the potter's wheel, wooden chariots, and spears fabricated from tin and copper? Or from even earlier Stone Age tales? Was Aristotle correct when he asserted that something within us "resists and opposes" rational principles?[255] We can collapse all such questions into a single inquiry: What is the source of religious belief? In answer, this chapter tenders seven theses. Each is selected from a larger body of research. Each is founded on a scientific footing. The first thesis comes from psychology.

RELIGION AS PSYCHOLOGICAL PALLIATIVE

In 1943 the American psychologist Abraham Maslow (1908-1970) published a paper on human motivation, setting the stage for the present-day palliative thesis.[256] In his paper, Maslow articulated a five-tier model of human needs which, for simplicity, I condense into two tiers: physiological needs (food, shelter, warmth, sleep) and psychological needs (security, belonging, love, fulfillment). According to Maslow, it is the desire to satisfy these needs that motivates many of our actions and behaviors.

Other psychologists, among them the American Steven Reiss (1947-2016),[257] extended the paradigm. Their research explicitly linked religious belief to the satisfaction of psychological needs. The connection is this: Our thirst to soothe psychological wants is strong, so strong we often grasp at any solution that promises to assuage our needs. Whether the promise is real is irrelevant. Whether any substance dwells behind the promise is equally immaterial. It is *belief* in the promise that is important. It is *belief* in the promise that satisfies our psychological needs. And it is this same *belief* - the overwhelming mental craving for the promise to be true - that attracts so many to Religion.

Religion not only understands this paradigm, but leverages it to profound effect. Simply put, Religion expressly designs its doctrines to palliate our psychological needs. It constructs its deities to inspire confidence and faith. It forges its doctrines to instill order and security. It structures its rituals to foster fellowship and interdependence. These and similar stratagems, each honed over millennia to near perfection, are highly effective. Consider a handful of examples related to our need for certainty, emotion, community, and meaning.

CERTAINTY Our species craves certainty, an attribute we can define as a combination of safety and predictability. Safety, of course, reflects our desire to be snug and secure, to be shielded from the uncertainties of life. Predictability expresses our hunger to know - to discern what lies ahead, to presage the future, and, where possible, to control it. Religion addresses both safety and predictability by providing clear, definite, and unchanging answers to pivotal questions. Is there a deity? Is there an afterlife? What does the deity want me to do or not do? What will happen if I obey or disobey? Sacred texts and sacred personages dispense practiced answers, removing myriad ambiguities and doubts in life.

Religion's favored technique for imparting certainty is the assertion of absolutes, an assemblage of truths that commonly extends throughout the dogma, jurisprudence, and moral philosophy of each faith. Each absolute is presented as clear and unwavering, to be relied upon in all circumstances. There is no true god but Allāh (Islām). Thou shalt not steal (Christianity). Practice right speech: speak truth, avoid slander, and refrain from abusive language (Buddhism). Avow non-violence (Jainism). The soul reincarnates (Hinduism). All men and women are equal (Sikhism). Absolutes underpin the ethical guideposts of nearly every faith: its distinctions between right and wrong, its contrasts between compulsory and forbidden acts, and its formulae for salvation. It is this assertion - this presentation of Religion's *what to think* body of knowledge as unequivocal, absolute, and immutable - that appeases our psychological need for certainty.

Notably, religious absolutes address what is likely the greatest apprehension of our species, the fear of death. To allay our anxiety, Religion paints death as a mere transitional phase, one that brings an end to the trials and sufferings of earthly existence. Beyond lies an ethereal realm where salvation and eternal life await, where virtue and righteousness abound, where loving family stand ready to embrace us - all under the auspices of the one true god. One must admit, it is a well-crafted, beguiling vision. Yet, here again, Religion falls into the trap of logical fallacy, the *appeal to the gallery* fallacy detailed in *The Case of Earth's Age*. The appeal, the reader will recall, plays to the sentiments of the audience, in this instance by alleviating a profound fear. Religion, untroubled by a complete lack of proof for its claim, refuses to contemplate a much simpler explanation. As long as our species fears death, we will invent supernatural entities for the express aim of soliciting their protection. In short, we demand the chimera.

Reason contends that Religion's ability to proclaim absolutes derives from the most rudimentary, quotidian explanation: Religion simply fabricates its truths, each free from the burden of verification. Reason, conversely, commands no similar prerogative. It demands that assertions be founded on credible, defendable, and testable evidence. Satisfying this demand requires methodology, experiment, expertise, diligence, and time, lots of time. Reason offers no quick fix. Moreover, Reason's *what to think* body of knowledge, by definition, is necessarily incomplete and open to revision. As such, Reason is less able to allay uncertainty. And while Religion offers simple and clear answers that nearly everyone can understand, Reason's techniques often complicate a topic or bewilder the layman: sample sizes, correlations, probabilities, confidence intervals, standard deviations, and multivariate analyses. All of which still do not produce a definite answer that is true in all cases.

Consider, too, a pair of complementary findings surrounding our need for certainty. First, it is no coincidence that polytheistic societies invented a deity for nearly every catastrophe that might befall them. Each offered safety from a specific threat. Monotheism simply consolidated equal, across-the-board protection into a single tri-omni deity - not unlike a present-day umbrella insurance policy. And second, note that our predilection for certainty often coincides with confirmation bias - the tendency to interpret information in a way that reconfirms our pre-existing instincts or beliefs. The same individual who accepts the certainty of a supreme being might see an image of their deity in a cloud or a piece of toast, interpreting the image as a sign that reinforces belief in the deity - a beautiful instance of pareidolia and its close cousin, circular reasoning.

The desire for safety and predictability extends, of course, beyond Religion and Reason. Politicians offer an everyday example. They appeal to the human need for certainty through the use of simple and understandable language, and by speaking with confidence and conviction: I will create jobs; I will reduce crime; I will end hunger; I will lead us to victory. Politicians steer clear of nuance and likelihoods, never uttering, "Our chance of success is 60%." They often favor gut instincts, eschewing facts and intricacies, even in the face of contrary evidence, to amass and motivate supporters. In politics as in Religion, Reason suggests a positive correlation between certainty and skepticism. The more assured the pronouncements, the more skeptical one should be.

EMOTION Religion's greatest strength lies in its ability to build emotional bonds with its followers. To those who despair, it accords hope. To those who grieve, it offers solace. To those who fear mortality, it promises an afterlife. To the guilty, it proffers forgiveness and redemption. To the unloved, it tenders a benevolent and caring deity. To the desperate, it sustains fortitude. To the distressed, it imparts peace. To the fatigued, it offers rebirth. To the blasé, it inspires awe. To its many adherents, it confers rewards in this life and beyond.

To nourish an emotional bond, it is no accident Religion instills its deities with human characteristics and desires: male authority figures, humanoid visages, longer lifespans, superior knowledge, and omnipotent power.[258] It simply is easier to identify with a deity that both resembles you and possesses characteristics you admire. It has been remarked that if giraffes and Baobab trees possessed belief systems and could draw, they would depict their deities as giraffes and Baobab trees.

In a similar vein, Religion often imbues its deities with universal human values - among them integrity, fairness, and virtue - to further cultivate an emotional appeal. As antithesis, juxtapose theism, the notion of a personal and caring god, with deism, which arose in the seventeenth century. Deism posits the existence of a non-personal god who created the world and the mechanisms that drive it. He or she then departed to attend to other tasks elsewhere. A common analogy is that of a clock-maker who constructs a clock and sets it running, then maintains no further involvement with the timepiece. Similarly, a deistic god is unaware you exist and never intervenes in worldly events. Unsurprisingly, the concept of a non-caring deistic divinity who offered no emotional bond fell from favor.

As we saw in *The Case of Human Origins*, external truth cannot be discerned by emotion. Yet Reason acknowledges emotion's unrivaled import to our species. It underlies our passions, our sentiments, and our affections. It motivates our thoughts and behaviors. It renders our greatest joys, our deepest sorrows. Its role in the survival of our species is unequivocal. It propels our selection of partners, our desire for children, our formation of friendships, and our joy of social interactions. For most of us, it grants the key to happiness. And simply put, it feels good. Arguably, no other aspect of our existence so profoundly touches the quality of our lives.

Reason cannot yet compete with Religion on emotion. Psychology, neuroscience, and related disciplines have not progressed to a degree sufficient to normatively measure and explain happiness, let alone recommend a reliable basis for public policy and individual action. Will Reason one day achieve this ability? I contend so, but that day is not imminent. At present, Reason is arguably dispassionate. It can extol the majesty of the universe and the beauty of a sunrise. It can marvel at the complexity of life and our ability to understand it. It can explain (or attempt to explain) love, friendship, and justice, but it cannot promise an ethereal, loving entity that watches over you and cares for you. This leads many to select a faith-based world of myth and mysticism rather than facing the sometimes chilly realities of Reason.

COMMUNITY As social animals, our species seeks and requires human interaction. Again, Religion adeptly offers this interaction by fostering a sense of community. Astute religious leaders cultivate fellowship at church assemblies, minister to the ill, befriend the isolated, visit new residents, lead youth programs, support community projects, address civic organizations, administer baptisms, counsel individuals on key decisions, and encourage church attendance. For those who attend church services, this act often satisfies a social longing as much as it signifies religious devotion, providing a regular opportunity to converse and cultivate friendships. Regular attendance further instills feelings of acceptance, forges social identity, encourages group cohesion, and fosters interdependence. These attributes acquire greater prominence in smaller communities where the church is the principal or sole social outlet.

Fear of exclusion is equally important. To those whose belief system represents a way of life, and for those where the church serves as the primary social outlet, exclusion from the group is unthinkable. The fear of ostracism drives acquiescence and conformity. Indeed, when the Catholic Church dominated medieval life in the West, excommunication often led to exclusion from the comity of community, the fear of which was no less than the loss of salvation.

Again, Reason struggles to compete with Religion on social interaction. Reason fosters friendships, even relationships, but less frequently provides the broad, all-embracing community of Religion.

MEANING It is the nature of our species to seek meaning. To find purpose, fulfillment, a grail of lasting significance. To adopt an organizing principle for our lives. To articulate a framework for our temporal existence. To be part of something transcendent. To discover "something more." Religion, ever astute, recognized this need from antiquity, constructing its scriptures and its teachings to proffer meaning. Consider the assertions of Christianity.

- We are proclaimed a special species: "God created man in his own image, in the image of God created he him; male and female created he them."[259]
- We are granted a special position: "Let them [humans] have dominion over the fish of the sea, and over the fowl of the air, and over the cattle, and over all the earth, and over every creeping thing that creepeth upon the earth."[260]
- We are offered happiness: "Delight thyself also in the Lord; and he shall give thee the desires of thine heart."[261]
- We are promised guidance:"In all thy ways acknowledge him, and he shall direct thy paths."[262]
- We are accorded salvation: "He that believeth on the Son hath everlasting life."[263]

Verses close in style tender guidance on wealth, power, vanity, pleasure, career, education, and sin. If we accept the deity and his teachings, we will find purpose and meaning. It is a powerful and enticing message, effectively delivered by Jesus, Muhammed, Guru Nanak (Sikhism), Yvāsa (Hinduism), and myriad other prophets and theologians.

Political and secular movements also recognize the need for meaning, often structuring their ideologies to satisfy the same psychological desires. So, too, do fringe religious movements. At intermittent historical moments, leaders emerge and leverage this inclination by proclaiming a new creed; by positioning themselves as the ones who know the truth; by articulating a "correct" way of living; by promising escape from an abhorrent status quo; and by offering hope, dignity, and future reward. Witness Karl Marx's Communist Manifesto, Mao Zedong's Little Red Book, and Al-Qaeda's doctrine of war against Jews and Crusaders. Each ideology, altruistic or nefarious, attracts hopeful and purblind followers, individuals willing to dedicate their lives to a truth proffered by the *prophète du jour*. Each truth endures until supplanted by a newer truth.

Do followers of Reason find meaning? Yes, but of a dissimilar stripe. Continuing in the vein of the prior discussion on

> emotion, Reason's practitioners find meaning in the grandeur of the cosmos, in the intricacy of nature, and in a world understanding based on fact. They derive purpose from the self-confidence to direct one's future and harness one's destiny, from the sublime "beauty is truth, truth beauty,"[264] and from the personal relationships that constitute the human experience. They know this life is the only life, and they value it accordingly. But Reason stops short, curbing the human desire to invent and deify an entity worthy of reverence. Unsurprisingly, the apostles of faith often find Reason's offer of meaning both impersonal and unfulfilling.

The number of human needs extends beyond the four just mentioned. Others include tranquility, esteem, order, justice, and vengeance. Note, however, there is no definitive list of the psychological needs of our species. Counts and categories vary by source, with most authors citing ten to twenty distinct needs. That said, the field of psychology, regardless of taxonomy, affirms a salient point. Yes, the attractiveness of Religion derives, in large part, from its ability to satisfy many of our psychological needs. But no, this ability in no way attests to the truth of religious claims. The palliative obtains its affect from *belief* in the promise, not from its veracity. Expressed in broader terms, psychologists recognize that the human mind is highly susceptible to want it wants to believe. So much so that, in pursuit of mental comfort, it often selects what it wants to be true rather than what can be shown to be true.

ALL-ENCOMPASSING EXPLANATION

As noted in *The Origin of Religion*, our species appears to possess an innate need to understand the world that surrounds us. Zaug and his tribe, unable to comprehend why the ground trembled, reflect the state of knowledge that prevailed for most of human history, where explanations of natural events were creative speculations, ambiguous fabrications, or simply non-existent. Reason could offer no assistance. It did not yet possess the tools and techniques to develop consistent, credible, and defendable answers. Religion, undaunted by its similar want of ability, filled the void with alacrity.

Religion tenders explanations for a broad spectrum of phenomena: puzzling natural events, vagaries of human health and happiness, origins of evil and suffering, and sensations of prescience and foreboding. In typical fashion, Religion presents its explanations as the acts of a deity, often in the context of a moral reward or punishment. We won the battle because we are pious and the deity was on our side. The flood that inundated our city resulted from impure thoughts. My brethren fell ill as punishment for disavowing their faith. Such visceral explanations endure today, affirmed by dissembling religious leaders of various denominations around the globe. A recent example was imparted by a Dominican friar, who ascribed an earthquake in Italy to God's punishment for gay civil unions. To its credit, the Vatican promptly and rightly decried the friar's comments.[265]

Note that religious explanations, regardless of faith, exhibit a ubiquitous pattern. They lack fastidious detail regarding the how, when, and why of any event. They cloak a systematic absence of empirical evidence. And they offer few, if any, falsifiable predictions. In Reason, such omissions are risible, even fatal. In Religion, they are habitual - an explanation's dearth of evidence is offset by its attribution to theistic authority. Yet, here, the endgame for Religion is intractable. As Reason advances, it not only becomes increasingly capable of deciphering the unknown aspects of our world. It likewise acquires the ability to refute all testable, faith-based affirmations. Religion, in contrast, never improves its ability to explain or predict events. Its only recourse is to ceaselessly rely on a litany of timeworn bromides, chimerical declarations, and evidence-free revelations.

MENTAL RELEASE

Religion, by dispensing theological frameworks and moral philosophies, offers yet another powerful inducement to its most devoted disciples: a sense of mental release. Prescribed guidance on a wide range of topics obviates considerable inquiry, analysis, and decision-making. The guidance and its sources are both absolute and time-tested. When considering a monthly tithe, consult the imām. When seeking absolution, consult the priest. When pondering the legitimacy of gay unions, consult scripture. When contemplating a course of action, pray. In this manner, the answer is revealed. Merely refer to ordained dogma, locate the germane precept, and behave accordingly.

Many religions go further by authoring a catechism, in the most general sense an exposition of the principles of their faith. The principles often appear in question-answer format, conveying a teacher-student or authority-apprentice relationship, with an expectation that the formulaic responses will be learned and cited by rote. Again, note the *Catechism of the Catholic Church*. Its 4 parts, 8 sections, 21 chapters, 67 articles, 234 sub-articles and 2,865 paragraphs[266] delineate the *what to think* (and how-to-behave) canon of the Church. With such guidance, myriad questions, nuances, and contradictory assertions simply disappear under the illuminating rays of faith. Harshly put, Religion reduces the need to think.

Reason, in contrast, demands critical thought, which manifests itself in two forms. First, by a dedication to learn the essentials of Reason's *how to think* process. And second, arguably a more onerous challenge, by sustaining the personal discipline to regularly apply its techniques. These include articulating insightful questions; considering competing viewpoints; weighing contradictory evidence; challenging deep-rooted assumptions; probing undisclosed motives; clarifying ambiguity; assessing pertinent information; emphasizing empirical evidence; isolating relevant variables; identifying patterns beyond immediate context; recognizing errors and faulty logic; engaging in personal reflection; and maintaining an awareness of one's biases. The object of all such techniques is to reach sound, reasoned judgments. The most crucial trait, however, is a willingness to accept uncomfortable conclusions (and the consequences of those conclusions) that might impugn long-held beliefs and convictions.

Fortunately, critical thought rarely requires knowledge of formal, academic reasoning systems. It is not necessary to learn associative and rules-based systems, propositional logic, the Hegelian dialectic, or comparable scholarly methodologies. Common techniques, such as the objective and skillful application of the six W's (who, what, why, where, when, how) normally suffice. Learning and applying Reason's *how to think* process nonetheless demands a commitment some choose not to make. Indeed, for most of us, thinking is hard. What school child prefers completing homework to watching a favorite video? What percentage of adults elect arduous study with no monetary reward? It is simpler to forego critical thinking, to be instructed *what to think*, to accept a mystical prescript, and to live one's life accordingly. It's simpler to believe a myth. Cue Religion.

As described earlier, Religion teaches *what to think* at an early age. When it subsequently injects its *how to think* methodology, the intent is solely to reinforce, not challenge, established dogma. Reason also often teaches *what to think* before *how to think*, but with a marked difference. Reason does not present its *what to think* canon as eternal truth. Instead, it presents its *how to think* process as a mechanism to contest existing truths and unearth new truths. This stands in stark contrast to Religion which, as we have seen, suppresses inquiry that challenges its *what to think* body of knowledge, precisely the type of inquiry that derives from critical reasoning.

Religion's discomfort with critical thought, in part, derives from the countless inconsistencies within its teachings, disparities that believers remain oblivious of or knowingly disregard.[267] If a deity knows our intentions, why is it necessary to pray to the deity? Why would a deity suspend the laws of nature to satisfy the entreaty of a solitary, self-avowed, undeserving supplicant? If a deity knows the future, why is the deity concerned how we act? In what enterprise was the deity engaged before the earth was formed? If a deity is tri-omni, how does the deity address his or her own boredom? Of the hundreds of billions of galaxies in the universe and their countless planets, why would a deity direct its attention to a single, self-important species inhabiting a solitary, rocky sphere? How does Religion respond to questions of this type? It fabricates. It obfuscates. It dissembles. And, in doing so, it enlists its staunchest allies: unsubstantiated claims, mystical revelations, and appeals to authority.

Religion's allergy to critical thought also promotes a complementary form of mental release, freedom from responsibility. If a faith declares that its deity placed mammals on Earth for the benefit of *H. sapiens*, the disciple need bear no accountability for the survival or extinction of each species. If the same faith teaches that its deity created Earth for the benefit of humankind, its tractable followers need accept no responsibility for climatic or ecological issues. Each individual, thus freed from obligation and censure, need take no action, need expend no energy, and need assume no guilt. The extinction of the Tasmanian Tiger, oceanic pollution, and global warming simply reflect "the will of the Lord."

Freedom from responsibility manifests itself in both implicit and explicit forms. The implicit variety derives from the behavior of the practitioner. Here I cite two examples. In 1900 U.S. Senator Albert Beveridge (1862-1927), in support of a bill to construct the Panama Canal, willingly relinquished personal responsibility when he informed other Senators that, "We will not renounce our part in the mission of our race, trustees under God, of the civilization of the world."[268] He went on to proclaim that God had been preparing the English-speaking and Teutonic peoples for this mission for a thousand years.

The English soldier and statesman, Herbert Edwardes (1819-1868), was no less resolute. He cited the defeat of the Sepoy Mutiny in 1857 as evidence of God's will to award India to the British for safeguarding the Christian religion.[269] History is replete with similar instances of individuals abdicating responsibility for their actions, and for all the repercussions from their actions, in the name of their deity who allegedly sanctioned the acts.

Explicit freedom from responsibility derives from theology, as attested by the French theologian, John Calvin (1509-1565). Second only to Luther in early Protestantism, in 1536 Calvin produced his masterwork, *The Institutes of the Christian Religion,* in which he systematically catalogs and explains the beliefs of the Protestant faith. What does Calvin contend in his multi-volume *tour de force*?

- Calvin proclaims the principle of predestination. He affirms that the God of Protestantism predetermines each individual's destiny. Free will and personal responsibility play no role in one's fate. In Calvin's words, "All are not created on equal terms, but some are preordained to eternal life, others to eternal damnation; and, accordingly, as each has been created for one or other of these ends, we say that he has been predestinated to life or to death."[270]
- He then declares that his dictum is confirmed by proofs from scripture, demonstrating yet again the clever use of circular logic: declaring scripture as truth, then arguing truth from scripture. Again, in Calvin's words, "This confirmation [is] founded on a careful exposition of our Savior's words, and passages in the writings of Paul."[271]
- To preclude rebuttal, Calvin divides all knowledge into two parts, both to be determined from scripture. He describes the division as "the knowledge of God, and … the knowledge of man."[272] Strangely, the latter refers to angels, devils, and the Biblical creation of the world. Reason plays no role in Calvin's world.
- He then cites authority to reaffirm his argument: "The mere fact of creation should lead us to acknowledge God."[273] His alleged "fact of creation" signifies his inability to imagine an evidence-based alternative to his deity-centric cosmos.

In one sense, Calvin's argument is cogent. A truly omniscient deity should indeed be aware of the actions and fate of each and every one of his subjects. Yet consider the consequences. Calvin's assertion of predestination grants a *carte blanche* dispensation from responsibility. If one's fate is predetermined, no action (or lack thereof) in this world can alter that fate. Attempts to shoulder additional commitment and responsibility will fail. Every decision, every act, can be interpreted as god's will. Whatever happens was predestined to happen. And to pose the by-now almost rhetorical question, what objective evidence does Calvin cite to defend his contentions? None. Religion once again articulates no hypotheses, offers no empirical evidence, and presents no testable predictions to support the claim of predestination. Accordingly, Calvin's assertions fall well below any bar of credibility.

To express what might not require expression, Reason rejects freedom from responsibility in both its forms, instead demanding individual and communal accountability.

CULTURAL CONSTRUCT

Psychology offers a number of explanations for religious belief beyond the three cited thus far - as psychological palliative, all-encompassing explanation, and mental release. Others relate to anxiety, guilt, deprivation, early childhood effects, and various forms of projection (ascribing attributes to others, while denying them in oneself). One of the more thought-provoking, however, stems from a merger between psychology and sociology. From broad research, the two fields offer some intuitive and not-so-intuitive findings:[274]

- Demographic and social variables drive religious inclination and its expression, more so than any other factors. Put another way, one's religious identity derives from the faith of their parents and their community. An individual's economic status, social status, education level, and gender also play a role. As such, election of faith rarely derives from an individual's neutral decision or from the truths of one particular creed.

- Specific religious experiences are instilled by culture. The content of such experiences, the situations in which they arise, and the types of people who experience them are predictable.
- Religious expression is a socially-trained construct requiring prior religious exposure. Without such exposure, religious behavior rarely exists.

These findings reveal two interrelated insights. First, individuals rarely choose their faith. Quite the opposite. Faith derives, in large part, from the prevailing beliefs of a particular geography and era. Consider the individual born in the West who devoutly believes in Jesus. That same individual, had she been born in northern Africa, would have expressed the same devotion to Muhammad or Allāh. Similarly, regard the present-day penitent reared in Punjab who avows the perfection of Ik Onkar. Had he resided in ancient Rome, he instead would have revered the sky god Jupiter. Thus, for the majority of individuals, the where and when of their birth determines religious belief. The prevailing deity plays no role.

The second insight is that Religion is cultural. It arises from the people who surround us.[275] Religious inclination, as well as the election of a particular faith, derive not from the will of a deity or from the truth of scripture. They simply arise from one's cultural environment. Change the culture and you change religious propensity. It's that straightforward.

We noted earlier that Religion explicitly designs its teachings and rituals to assuage our psychological needs. The same holds true for culture. Religion, comprehending that faith derives from place of birth and social context, takes steps to ensure that its beliefs, and its beliefs alone, reach the ears of its followers. Among the myriad tools at Religion's disposal, we will mention two: indoctrination and manipulation.

INDOCTRINATION Indoctrination, a mainstay of Religion, deliberately begins at a young age, well before children develop critical reasoning skills. Religion instructs toddlers that an invisible entity watches over them. It informs them how the unseen entity expects them to behave. It entrances them with superhuman, magic-like tales. It entertains them with religiously-themed games. It coaches them to recite godly phrases. It trains them to adopt physical postures in accordance with the dogma: kneeling, bowing the head, closing the eyes, tracing the sign of a cross, or aligning their bodies to religiously significant compass coordinates. It warns innocent, uncritical minds that future happiness arises only from acceptance of the one true god. It likewise threatens divine punishment should the unwise child fail to obey. To embed its precepts, Religion harnesses the foremost instrument of indoctrination, repetition. Repetition, repetition, repetition. Followed by still more repetition. Incessant repetition. Repetition choreographed to instill an ingrained, unquestioned conviction from early age through adulthood.

Two of countless examples evince the point. In Judaism, the coming-of-age rite of Bar Mitzvah occurs at age 13 for boys. The equivalent rite of Bat Mitzvah occurs at age 12 for girls. The two rites signify full entrance into the faith, with the encompassing rights and duties. Note the timing of the rituals. Both occur at the apex of early childhood inculcation, just as youth attain the capacity for critical thought. A less well-known but indicative case is the 4/14 Movement, whose purpose is to spread the word of Christ to children worldwide in the 4-to-14 age demographic.[276] Again, note the targeted age range. In its proselytizing materials, the Movement cites scripture to underpin its focus on youth, seeking to establish a "stronghold from the mouths of children."[277] Religion understands that capturing the minds of the young often creates disciples for life.

Large-scale indoctrination requires large-scale organization. Religion possesses both the means and method. It indoctrinates through its clergy, its churches, its religious schools, its missionaries, its doctrines, its prayers, its catechisms, its charities, its symbolism, its moral codes, its rites of passage, its festivals, its feasts, its music, and its art. Consider a ubiquitous example: weekly church attendance. An authority figure calls the penitents to order. Congregants are instructed when to sit, when to stand, when to sing, when to be silent, when to open a sacred text, when to recite a holy passage, when to bow their heads, and when to pray. There is no diversity of thought, no questioning of authority, no deviation from the formula. Religion, admirably proficient in its rituals, conducts them expertly and flawlessly, each burnished over millennia to a fine luster.

To reinforce the indoctrination, Religion extends its partisan efforts with fantastical, faith-based stories. Again, to cite one of endless instances: In Mexico, the farmer Juan Diego Cuauhtlatoatzin (1474-1548) reported seeing and conversing with the Virgin Mary. The Virgin requested a chapel be constructed and, as proof of her petition, imprinted her image on Juan's cloak. A chapel duly was built housing the cloak and its miracle image, later replaced by a basilica. Today, a newer basilica houses the garment, which is an object of pilgrimage across Latin America. Pilgrims walk for miles (or crawl on their knees) to the basilica, soliciting blessings and cures from the Virgin. The act inspires both the humble penitent and the faithful spectator. The Virgin Guadalupe, aka the Virgin Mary, became the patron saint of New Spain (present-day Mexico and surrounding areas) in 1754. The Catholic Church, ever adept at thaumatology, canonized Juan Diego in 2002.[278]

Adherents to the faith, themselves subjects of indoctrination, extend the edification within their homes, within their social circle, and, often, within their workplace. In communities where a majority of the inhabitants embraces a common spiritual ethos, the community itself becomes a reinforcement mechanism. All members are expected to adhere to the religiously-based cultural norms. The same occurs when a nation state aligns itself to a particular belief system, transforming all its apparatuses into reinforcement mechanisms - its leaders and administrators, its politics and policies, and its laws and regulations. In Saudi Arabia, for example, the government mandates religious instruction in public schools and requires all citizens be Muslims. For many believers in these societies, there is no separation between religion and secular aspects of life. Indeed, citizens may be unable to envision a separation.

Individuals, once indoctrinated, tend to remain within the belief system of their birth. Exploration of faiths outside the culture often meets with disapproval, even punishment. Personal guilt is an equally potent deterrent. An individual who contests ingrained beliefs often abandons inquiry amid feelings of excessive guilt.

Indoctrination also leads highly-religious communities to regard belief as a credential, where politicians, teachers, physicians, electricians, plumbers, and other professionals have no choice but to cite adherence to the prevailing belief system as evidence of their suitability to perform secular roles. Perhaps a Jewish mechanic is favored over a gentile in Jerusalem. Perhaps a Buddhist policeman is desired over an agnostic in Bangkok. Perhaps a Shia accountant is preferred over a Sunni in Baghdād. Politicians campaigning for office in the United States, for example, often eschew Reason, relying instead on demonstrations of religious belief to garner votes.

In some societies, individuals who fail to profess the "correct" religious credentials become targets. Indonesia provides a recent example, where the Christian governor of Jakarta was accused of misusing a verse from the Qur'ān and insulting Islām. [279] A hardline Islāmic minority in the predominantly Muslim capital doctored a video and fabricated accusations to eliminate the non-Muslim candidate, who was tried and convicted of blasphemy.[280]

It is important to reiterate that Religion inculcates its *what to think* canon long before instructing *how to think*. As such, the subsequent application of the five sacreds for teaching *how to think* simply becomes a reinforcement mechanism; the methodology in no way inspires a search for alternative truths or enlightened worldviews. There might be moments in history where Reason can be accused of indoctrination but, where it exists, it is short-lived. Indoctrination is not a fundamental tool of Reason. Reason obliges no four-year-old child to bow its head in reverence to the laws of gravity, to prostrate itself to wave-particle duality, or to intone a hymn in praise of DNA's double helix. Reason never asserts one and only one immutable truth. The spirit of Reason is to pursue the path of experiment and evidence, wherever that path might lead.

MANIPULATION Manipulation shares attributes with indoctrination, but its motivation differs. For our narrative, we define indoctrination as promoting a doctrine reputed to be true, whereas manipulation willfully advocates a patently false ideology. One form of manipulation is the noble lie[281], a stratagem articulated by Plato (c.428-c.348 BCE) in *The Republic*. The artifice employs a fiction, typically a religious myth, to sustain the hegemony of a ruler, maintain social order in a society, and regulate the behavior of its citizens. Plato asserts that the "fostering of such a belief will make them [the people] care more for the city and for one another."[282] He further acknowledges that the sole aim of the myth is to prepare the people to accept the laws of the rulers. He invokes a metaphor, one of applying permanent dye to cloth, when he states, "we were contriving influences which would prepare them [the populace] to take the dye of the laws in perfection."[283]

The English historian Edward Gibbon (1737-1794) recorded a similar use of the noble lie in the polytheistic society of ancient Rome: "The various modes of worship ... were all considered by the people, as equally true; by the philosopher, as equally false; and by the magistrate, as equally useful."[284]

Late eighteenth-century France provides yet another example. The French Revolution, among its myriad goals, attempted to introduce a secular state. Toward this end, revolutionary leaders limited the power of the Catholic Church by nationalizing its property and disbanding its religious orders. Napoleon, however, recognized the importance of religious practice as a means of establishing his dominion over the political and social affairs of the nation. Upon his rise to power, he reached an agreement with Rome, titled the *Concordat of 1801*, to reinstate the Church in France. The compact transferred authority to the French state, granting it a free hand to appoint bishops and to administer church finances, thus awarding Napoleon his intended instruments of control.

The aura of secrecy surrounding a noble lie makes it difficult to ascertain the extent of its present-day use. Its application likely continues at different times in different places, orchestrated at various levels of secular and religious hierarchies. Its rationales remain the same as in Plato's age: to validate the legitimacy of a ruler, to persuade a population to accept its lot in life, or to promote a perceived societal good. The authority of elites drives acceptance of the lie. Its implementation is guided by the mechanisms of the state: from incorporation into its laws, by reinforcement through its customs, and through the education of its children. The consequence is that the citizenry believes because it has been taught to believe. In this context, faith is a useful, pious fiction.

PALEOLITHIC ADAPTATION

Psychological and sociological explanations for religious belief, while insightful, at times feel self-evident. Not because they are cursory or unexamined. To the contrary, considerable research lies behind the explanations. Yet on their own, they fail to pinpoint an underlying cause. Why is our species susceptive to religious belief? Why do we allow ourselves to accept faith-based claims with no supporting evidence? Why do we prefer myths that satisfy our psychological needs rather than fact-based explanations that might not? Several disciplines - among them evolutionary biology, evolutionary psychology, and neuroscience - attempt to examine religious belief at this more fundamental level. Their conjectures, in one form or another, relate to the brain.

Our brains afford us a capability for abstract thought and a competency for language - not a limited set of grunts and calls, but a rich and extensive competence to communicate both verbally and non-verbally. *H. sapiens* and (possibly) our extinct hominin relatives are the only species to have evolved a brain with these capabilities. Unsurprisingly, our cerebral prowess underlies our ability to comprehend both Religion and Reason. But evolutionary psychology, basing its ideas on natural selection, suggests differing predilections for Religion and Reason. The story, briefly put, is as follows.

Perhaps some of our paleo-ancestors were better able to detect sources of harm. Perhaps this ability arose from the limbic system, a region of the brain concerned with instinct and basic emotions (anger, love, trepidation, dominance). Perhaps individuals acting on instinct were better able to escape danger - when a sound in the brush threatened injury, when a movement in the forest signaled peril, when darkness obscured unknown risks. Perhaps, in the same way, subtleties in the limbic system interpreted uncertainty itself as a threat. Perhaps a desire to remove uncertainty set the stage for spiritual inclination by ascribing the cause of natural events to the agency of unseen forces.

Evolutionary psychology suggests that traits of this type offered a survival advantage in early hominin societies. The survivors were more likely to reproduce and transmit their genes to future generations. This does not imply a gene for Religion, but it suggests that evolution selected emotion-based behaviors for their survival advantages. Religion, itself an emotion-based behavior, hitched along for the ride. In this scenario, religious inclination is a consequence of a brain evolutionarily biased toward emotion-based affinities. So in lieu of asking whether Religion encourages belief with no evidence, the more appropriate question might be whether evolution produced an "emotion heavy, evidence light" human brain that makes *H. sapiens* susceptible to belief with no evidence.[285]

This explanation also offers insights on Reason. Emotion-based and perception-based impressions enable speedy interpretations of our environment with nearly effortless thinking. Hard-coded, friend-or-foe, fight-or-flight responses require only facile appraisal.

Critical thought, by contrast, with its emphasis on disciplined analysis and objective evaluation, is comparatively lethargic, perhaps offering a less immediate survival advantage in early hominin societies. If the evolution of our brains favored emotion-based traits as a survival strategy, this might explain why critical thinking is a learned skill.

NEUROCHEMICAL REACTION

Neuroscience offers a related brain-based thesis, suggesting that chemical molecules and neural connections in the brain indirectly promote religious belief. The idea is that certain behaviors and experiences induce the brain to release serotonin, dopamine, and other neurotransmitters that, in turn, invoke feelings of well-being. Religion and its rituals elicit similar feelings, suggesting belief is a function of the make-up of the brain and its neurochemistry. This explanation creates a self-reinforcing framework. Religious experiences release neurotransmitters; neurotransmitters invoke feelings of well-being; feelings of well-being stimulate further religious participation; further religious participation releases more neurotransmitters, ad infinitum.

Experiments with psychedelic drugs lend support to the idea of neurochemistry, where participants receiving the drugs reported an increase in religious and mystical experiences. Here, Jim Morrison comes to mind. Other experiments concluded that meditation, music, and religious language also enhance feelings of unity, transcendence, and sacredness in test subjects, possibly from similar neurochemical causes.[286] Still other studies noted that, during prayer, blood flow diminishes to the area of the brain responsible for complex behaviors. At the same time, activity spikes in the limbic system, the brain's emotional center.[287] At present, none of these results is conclusive, but neuroscience clearly offers more to learn.

EVOLUTIONARY BY-PRODUCT

Evolutionary explanations for religious belief tend to focus on adaptive values, that is, useful survival attributes. I choose to include as the seventh and final thesis in this chapter an explanation that, at least on the surface, offers no adaptive value - that of Religion as a simple by-product. The trait to be examined is nipples.

In females, nipples are functional. In males, they are not. So why do males have nipples? Erasmus Darwin believed in the inheritance of useful traits, a form of Lamarckianism discussed in *The Case of Human Origins*.[288] This led Erasmus to surmise that male nipples once were functional. Perhaps males of some species lactated, perhaps males and females were shaped independently by natural selection, or perhaps both males and females evolved from some common hermaphroditic forebear. These ideas, while creative, proved to be dead ends.

Gould suggested a more likely scenario. Males and females start their development from the same type of embryo, with gender-specific instructions encoded in DNA. As the embryo grows, its genetic code directs the emergence of body parts fitted to each gender. Consider the volume knob on a Bluetooth speaker as a metaphor. Turn the knob to the left for a softer sound, to the right for a louder sound. Now, imagine a similar nipple knob. If the embryo is male, the DNA code turns the nipple knob to the left, producing smaller, non-functional nipples. If the embryo is female, the DNA code turns the knob to the right, producing larger, functional nipples. Thus, the reason males have nipples is it's simpler to produce them than not to. It's simpler to direct the epicene growth of the embryo from a single design, from a single set of DNA code. As Gould said, males have nipples because females need them.[289] In the same vein, a proclivity toward religious belief might simply be a by-product of other, emotion-based evolutionary adaptations in the brain, serving no adaptive purpose on its own.

From what evolutionary adaptations might religious inclination have arisen as a by-product? Ideas abound, but as yet there is no clear answer. It is just as true that evolutionary theory cannot yet definitively explain the origin of religious belief. But moving beyond self-evident explanations in search of underlying causes is a step in the right direction. This is a task for Reason, as Religion is unlikely to perform an exhaustive self-examination. It possesses neither the ability nor the will. Moreover, some faiths interpret such inquiries as attacks by iconoclasts on orthodox teachings.

FROM CAUSES TO CONDUCT

Reason, in its search for knowledge, is and should be indifferent to existing beliefs. It does not matter what you believe, just as it does not matter what I believe. What matters is what can be demonstrated via Reason's *how to think* tool kit. This provides cold comfort to individuals who see Reason as a threat to their faith-inspired world. Religion's instinctual response is not to quietly acquiesce, but often to denounce Reason and its discoveries, raising the question whether Reason can function in highly-religious cultures.

The short answer is no, it cannot. It might be permitted to operate in fields that produce new technologies, but not in the broader sphere of scientific inquiry to understand the universe and all it contains. Conservative religions, in fact, often construct the most intolerant cultures, those least friendly to inquiry. Their literal interpretations of scripture demand unquestioned acceptance. It is no accident Reason floundered during the oft-intolerant Middle Ages in Europe. It was not from want of technology. The Greeks were a beacon of Reason a millenium earlier with the same or less technology. War and instability played a role, as Reason requires stability to advance. But the dominance of religious authority in nearly all aspects of life was the primary antagonist, effectively thwarting and punishing unwelcome inquiry. If one believes, that is, is taught to believe, that all truth derives from a deity, that all inquiries can be answered by a trite *Goddidit,* there simply is no need to search elsewhere.

The same state of affairs exists in parts of today's world. Al-Khalili cites that the Muslim world spends less than one-fifth as much on research and development than the developed world. He further cites that Muslim countries have fewer than ten scientists, engineers, and technicians per 1,000 of population, less than one-half as many as the developed world, and that they contribute less than one percent of the world's scientific papers.[290] Approximately 60% of university graduates in Saudi Arabia, for example, earn degrees in Islāmic subjects. And the few who graduate in scientific fields must practice their profession in a decidedly inquiry-unfriendly environment, where criticism (or the perception of criticism) of Islām, the ruling family, and the government is forbidden. A broad definition of "crime" allows the government to suppress any unwelcome behavior or thought, among which: advocating equal treatment of Shia Muslims in a Sunni society, committing blasphemy, expressing atheist thought, mocking religious symbols, and engaging in immoral activity. In short, anything that could be interpreted as a threat to Islāmic values. Punishment of convicted perpetrators encompasses detainment, harassment, discrimination, prison, torture, or execution. Saudi Arabia is a severe case, but throughout much of the Muslim world, the contemporary relationship with Reason differs widely from the association thought to have prevailed during the Golden Age of Islām.

Religion's compulsion to constrain Reason is not limited, of course, to the Muslim world. The examples cited throughout this monograph testify that Christianity often pursues a similar aim. Nor is the pursuit restricted to the largest religions. In Israel, for instance, Haredi Jews (Ultra-Orthodox Jews) comprise the fastest growing segment of the population, currently at 14%. Their educational focus for males? An almost single-minded preoccupation on religious studies, punctuated by an unabridged absence of science and other secular subjects. Their foremost political motivation, of course, is to influence the state and society toward a more religious direction.

In America and parts of Europe, supporters of creationism continue to press a related agenda. Their politically astute members seek to control school boards and the content of textbooks, all under the guise of protecting Religion and forcing the teaching of their religious beliefs as legitimate science. Others enroll their children in private religious academies or home-school their offspring, curtailing access to Reason during the all-important, formative years of religious indoctrination.

In a relatively recent instance, the State of Arkansas passed the *Balanced Treatment for Creation-Science and Evolution-Science Act*, requiring public schools to teach religiously-inspired creation stories as science. Creationists hold the earth is, at most, 10,000 years old; all organisms were created in their final, non-changing forms; and the fossil record was deposited in the Biblical flood. A lawsuit challenging the Act was successful, and in 1982 the Act was declared unconstitutional. The judge in the case, William Overton (1939-1987) ruled that creation science does not employ scientific methodology, does not present testable and falsifiable assertions, and is merely a bid to promote a specific set of religious beliefs.[291] The judgment further dismissed the creationist claim that scientific evidence which fails to support evolution, by that very fact, supports creationism (the *God of the Gaps* argument). To the credit of Religion, a number of the plaintiffs (those seeking to strike down the law) were Methodist, Presbyterian, and Jewish religious leaders.

In a 1987 case regarding a creation science law in Louisiana, *Edwards v Aguillard*, the United States Supreme court struck down the law as unconstitutional for similar reasons.[292] In support of the defendant, 72 Nobel prize-winning scientists and 17 state academies of science filed briefs against creation science.

In 2005 a Federal court struck down a school board policy in Pennsylvania that, yet again, sought to advance creationism as science. In *Kitzmiller v Dover Area School District*, Federal court judge John Jones III (1955-) struck down the policy, re-confirming that intelligent design (a rebranded form of creationism) is a thinly veiled attempt to promote an express set of religious beliefs and is not a scientific theory. The judge, in reference to "the breathtaking inanity of the [School] Board's decision," noted that, "It is ironic that several of these individuals, who so staunchly and proudly touted their religious convictions in public, would time and again lie to cover their tracks and disguise the real purpose behind the ID [Intelligent Design] Policy."[293]

Permit me a pause at this juncture. Creationism warrants further commentary. I find it curious that creationists often invoke the vocabulary of Reason, employing terms like *theory* and *evidence* to serve their ends. When creationists utter the word *theory,* for instance, they seek to elevate their dogmatic claims to the objective standard of Reason. Their act is reminiscent of the parable of the vexed potter who refuses to discard a valued creation cracked in firing. Instead, he fills the cracks with wax, then covers the entirety with a glossy varnish - all to disguise its imperfections and convey superficial perfection. The equivalent verbal act allows creationists to contend that their theories should receive equal and impartial emphasis in public education. The fallacy of this stance is manifold, but I limit the narrative to three of the more egregious deceptions.

First and foremost, as the preceding court cases demonstrated, creationist theories are crafted to ratify specific belief systems - the well-worn practice of natural theologians to mimic the superficial veneer of the potter's fakery. On no account do they bespeak an objective search for knowledge.

Second, the theories manipulate scientific data. As stated in *The Origin of Reason,* no law is correct 100% of the time. Take tides. Tidal behavior is well understood. Diurnal (one high and low tide per day), semi-diurnal (two high and low tides per day), and mixed (uneven) tides derive from the gravitational pull of the sun and moon, and from the rotational properties of the earth and moon.[294] Yet predictions of specific tidal action within the irregular landmass of a queerish cove might fail. There are always anomalies. Creationism cleverly identifies the anomalies, then sculpts its theories in alignment to its dogmatic ends. A gap in the fossil record, an inconsistency in radiometric dating, an unexplained fluctuation in gravity, a disagreement among evolutionary biologists - all serve creationism's dodge. The clever packaging of a cherry-picked set of such aberrations conveys a sense of credibility to the casual observer. Yet creationist scholarship is notable chiefly from its absence. Creationism generates little of its own evidence, preferring instead to critique the evidence of real science. Moreover, its theories invariably lack the objectivity and rigor of Reason. Let's remind ourselves that a theory is a well-substantiated explanation founded on a comprehensive body of evidence, offering testable predictions that explain a broad range of phenomena and correctly predict the outcomes of new observations. Creationist assertions fall well below this bar. As the courts confirmed, creationism's use of the word *theory* is merely a pretense. Were it not, anyone with a passing hunch could exploit scientific flukes, ignore the dictum requiring a preponderance of evidence, label the hunch a theory, and clamor for its instruction in public education - all to serve a desired religious or political agenda.

Third, to once again re-affirm the findings of the courts, creationism not only disregards the rigor of Reason, but shuns any pretense of falsifiable predictions. When creationism's predictions are disproved, its proponents never acknowledge the absurdity of their claims. They never admit the game is up.

If highly-religious cultures impede Reason, what about specific individuals living within these cultures? Do the distinct *how to think* processes of Religion and Reason prevent followers of the former from contributing to the latter? It depends. On the one hand, history is awash with examples of religious individuals who contributed to Reason. Isaac Newton (1643-1727) was a devout Anglican. The Belgian priest George Lemaître (1894-1966) was a brilliant mathematician who proposed the novel idea of an expanding universe. The Muslim scholar Al-Idrīsi (1100-1165) was the greatest cartographer of his age. The Muslim physician Ibn al-Nafīs (1213-1288) described the circulation of blood. The English Quaker John Dalton (1766-1844) was an early proponent of the atomic theory of chemistry. Yet it sometimes is difficult to uncover their true motivations. The English polymath Robert Hooke (1635-1703), for example, attributed his microscopic discoveries, his automatons, to the "All-wise God of

Nature."[295] Whether he wanted to acknowledge a deity or not, he had to. In cultures where Religion dominated, natural philosophers often adopted silence, discretion, and other protective behaviors to avoid accusations of heresy or apostasy.

In today's world, the deciding factors are twofold. First, is the individual willing to ask and investigate questions which might contradict their faith? And second, is the same individual willing to accept the objective results of their inquiries, outcomes which might repudiate their faith? It is easy to identify such individuals if one can obtain an honest answer to a single question: What would it take to alter your beliefs? The response will fall into one of two categories: either a declaration of faith which no measure of reasonable evidence could alter or, conversely, an affirmation that education and new discoveries could inaugurate a fresh perspective. Those who cling to the former response find refutation of their faith inconceivable, are unwilling to challenge their own deeply held convictions, and lack the courage to follow objective evidence wherever it might lead. They conjure a mental perimeter beyond which there is no trespass, thereby limiting their potential contributions to Reason.

The question whether highly-religious individuals can contribute to Reason raises a more general inquiry. Is it possible to alter an individual's beliefs? Once youth attains maturity, is the adult able to recast a deeply held conviction? Psychology, once again, offers a number of insights on this topic. One relates to cognitive dissonance, a proposition developed by the American psychologist Leon Festinger (1919-1989).

Cognitive dissonance refers to the psychological discomfort individuals experience when they are forced to confront ideas or events that contradict their own beliefs, opinions, attitudes, or values.[296] The mental discomfort motivates the individual to reduce or eliminate the dissonance. How is this achieved? Consider three individuals: an idler who tires of suggestions to initiate a fitness regime; a staunch conservative who scorns the election of a liberal leader; and the driver of a fuel-inefficient vehicle who dismisses climate change.

When faced with information that contradicts their values, each individual has two choices, *adopt* or *deny*. The individual who selects *adopt* evaluates the validity of the contradictory information and, when warranted, alters her beliefs and behaviors accordingly. The idler exercises, the conservative acknowledges the merit of some liberal policies, the driver purchases an electric car. The individual who selects *deny*, however, rejects or ignores the information and perseveres with his previously-held conviction. The idler rationalizes that some unfit people live to an old age, the conservative condones dubious legislative tactics to block the liberal agenda, the driver concludes global warming is a hoax. Both choices, *adopt* and *deny*, restore the individual's cognitive consonance - their mental tranquility.[297] One's beliefs once again align to one's perception of reality. All is well.

Few like their beliefs challenged, yet not everyone experiences cognitive dissonance to the same degree. Who is more likely to suffer dissonance? Who is more likely to alter their beliefs when confronted by contradictory information? Studies suggest distinctive patterns.

- Introverts are more likely to suffer from cognitive dissonance, a proneness that perhaps derives from a greater propensity for self-reflection. Yet introverts also are more likely to change their minds when presented with relevant new information.
- Extroverts are less prone to suffer, and they are equally less inclined to change their minds when confronted by facts that refute their convictions.
- Highly confidant individuals also suffer less, but also are less likely to be correct in their assertions. Perhaps such individuals focus more on portrayals of confidence than understanding (or caring to understand) the salient and granular elements of a topic.

- Individuals who seek a higher degree of certainty in their lives suffer more.
- Younger individuals experience less dissonance than older individuals.
- Individuals who do not expect to encounter dissonance, and who thus are less defensively prepared when contradictory information arises, are more likely to consider new information.
- By and large, the more strongly held the conviction, the greater the sensation of cognitive dissonance when contradictions arise. Equally, if a contradiction challenges not one but a cluster of convictions, dissonance increases.

Those who select *deny* sometimes exhibit a more extreme behavior - an unwillingness even to consider contradictory information. Their mental reaction is not simply *I don't believe it.* It extends to *I don't want to know, and I don't want to think about it.* An individual susceptible to this mindset actively avoids situations that create dissonance. He maintains consonance by selecting only like-minded social groups, by culling incompatible news and information sources, and by reinterpreting inconvenient facts.

And it extends still further. The refutation of a deeply held conviction can strengthen an individual's belief in the "truth" of the conviction. Let's repeat this non-intuitive finding. Disproof can harden commitment to an ingrained conviction. A well-documented instance concerns the Seekers, a UFO (Unidentified Flying Object) religion from the 1950s. The religion's originator, an Illinois housewife named Marian Keech (a pseudonym), claimed to receive messages from the extrasolar planets Clarion and Cerus. Their denizens, the Guardians, communicated to Ms. Keech via automatic writing - the professed psychic ability to transcribe the words of a spiritual or supernatural entity, with no conscious effort or awareness from the writer. In the 1940s Ms. Keech began receiving the messages, which she shared with a small group of uncritical followers, the Seekers. The Guardians imparted wisdom from their home worlds and established a relationship with the Seekers. Among their messages, the Guardians asserted that human ignorance of universal laws was the cause of misery on Earth.

The messages took a darker turn in 1954, culminating in a prophecy of worldwide upheaval. The Guardians heralded the end of the world on 21 December of the same year. Among their prognostications: A flood would inundate the central lands of the United States, Canada, and Mexico; a new mountain range would rise in the eastern USA to contain the flood; France, England, and Russia would sink to the bottom of the sea; Egypt would be reformed into a fertile valley; many would perish; the resurrected would be taken up in a burst of light; Earth would be purified. The Seekers dutifully gathered on the evening of 20 December to await the end. They were to be saved by promised evacuation in Guardian spaceships. Their hours-long vigil terminated in disappointment, but not disbelief. Despite the unarguable refutation of the prediction, the Seekers retained and re-affirmed their beliefs.

Festinger and his colleagues (fellow psychologists) studied the Seekers. One goal was to understand, in Festinger's words, "How and why does such a response to contradictory evidence come about?" They identified five conditions under which one can "expect to see increased fervor following the disconfirmation of a belief."[298] They are:

- A belief must be held with deep conviction.
- A person holding a deep conviction must have taken some action that confirms their commitment to the conviction (publicly proclaimed a belief, resigned from a job, separated from a spouse, donated a sizable amount of money).
- The belief (or prediction) must be sufficiently specific and "real world" so that it can be refuted.
- The refutation must be recognized by the holder of the belief.
- The person must have social support that reinforces the correctness of the belief.

The case of the Seekers satisfied all five conditions. The upshot? They elected to eliminate cognitive dissonance by clinging to ingrained convictions in the face of repeated disproofs, refutations which even they acknowledged. It could be argued that the Seekers provide an extreme example of those who select *deny* to restore cognitive consonance. Yet their case is not uncommon. History relates scores of similar end-of-the-world prognostications by Religion, plus innumerable other disproved predictions. The refutations all too often provoke a Seeker-like reaction: the elevation of faith above Reason, an artifice whose sole purpose is to preserve a mythical ideology.

To cite a final example from the Seekers, the Guardians issued a warning against scientists and unbelievers, claiming that both factions are led by Lucifer against those who follow the light under the banner of God. The Bible, Qur'ān, and other scriptures trumpet similar warnings. "Beware of false prophets."[299] Only "Allāh guides to the truth."[300] Zealots and ideologues then echo the warnings with handy admonitions: Beware this test of faith, heed the word of the Lord, and shun the work of Satan. Again, all such banalities merely serve to dispel distasteful evidence and restore consonance among the most devout.

With the above insights, let's return to the original question. Is it possible to alter one's beliefs? The field of cognitive dissonance suggests it is difficult, but not impossible. The personality of the individual comes into play, as does the strength of the conviction. Reasoned argument, message repetition, and the authority (and attractiveness) of the message bearer also play a role. Of greater import, many suggest, are the convictions of the social group. An individual exposed to a new, dissonance-creating idea is less likely to select *adopt* if their social group rejects the idea. In Religion, as we have seen, the influence of the social group is strong. Deviating from the group mindset might expose the individual to censure, ridicule, or ostracism. For the person striving to adopt a new belief against the norm of the group, the answer is harsh but pragmatic. The individual often must seek a new social group whose outlook aligns to the new information. The new group then acts to restore consonance for the enlightened individual.

Some sources answer this question in context of *belief-in* versus *belief-that*, perhaps to separate cognitive dissonance in the religious and secular spheres. They first define *belief* simply as a mental attitude that something is likely to be true. They then designate *belief-in* as belief in someone or something transcendent, a non-empirical dogmatism characteristic of religious thought. *Belief-that* denotes a temporal belief supported by evidence that can be verified or refuted. Unsurprisingly, proponents of this distinction note it is simpler to alter *belief-that* viewpoints than *belief-in* convictions. In other words, Reason has greater success among those who favor a fact-based world.

The bottom line? Cognitive dissonance, as a psychological aspect of our brains, motivates us to seek and maintain consonance. It, among other factors, drives the unsatisfactory but realistic answer to our question. Attempts to change one's beliefs are not always successful. But one can try. Start young.

Let's close this section with some statistics on believers and non-believers. Take the statistics with a grain of salt. No systematic collection of global data exists on belief, particularly on non-belief. And there is no common definition of non-belief. Yet here are some indicators. Non-believers tend to be younger, male, more educated, more affluent, less authoritarian, and less suggestible. Believers, the opposite. Those who tick the "I don't believe in God" box on surveys represent 1% to 24% of a nation's population. The percentage is lowest in Venezuela and the Philippines, highest in Germany and France. There are at least 500 million atheists and agnostics worldwide. The share of non-believers has increased in recent decades. In the United States, non-believers are five times more likely to describe themselves as Independents than Republicans, and four times more likely to classify themselves as Democrats than Republicans.[301]

In this chapter, we explored a number of psychological, sociological, and evolutionary explanations for religious belief. The explanations suggest four answers to the question posed at the outset of the chapter: Why do people believe in ancient tales of divine phenomena? The responses, tersely encapsulated, are as follows:

- The individual, believing their decision is freely chosen, declares: "I believe because I choose to."
- Psychology, from its examination of our psychological needs, confirms: "I believe because it makes me feel good."
- Sociology, from its studies of demographic and social factors, affirms: "I believe because I have been taught to."
- Evolution, from the make-up of our brains as forged by natural selection, asserts: "I believe because my brain predisposes me to."
- Yet in no case do empirical evidence and testable predictions suggest: "I believe because the tales are true."

We also explored the ability of Reason to operate in highly religious cultures, arriving at a mostly negative conclusion. This is a pattern of history, and it continues to this day. At different times in different places, Religion musters its forces of natural theology - the suppression of rational inquiry, the dismissal of evidence, and the avowal of not-testable assertions - to maintain its mystical beliefs. And it continues to employ its primary weapon, invocation to higher authority, to justify both its doctrines and its acts. For its own survival, it has no other choice.

MISCELLANY

The number of paragraphs in the *Catechism of the Catholic Church* is constant at 2,865, but their organization into chapters, articles, and sub-articles varies by source.

At the time of this writing, the state constitutions of all 50 of the United States refer to god or a divine being. At least seven cite belief in a supreme being as a prerequisite for holding public office. The seven appear below. Enforcement of the statutes is unclear.

> Arkansas Article 19, Section 1: No person who denies the being of a God shall hold any office in the civil departments of this State, nor be competent to testify as a witness in any Court.
>
> Maryland Declaration of Rights, Article 37: That no religious test ought ever to be required as a qualification for any office of profit or trust in this State, other than a declaration of belief in the existence of God.
>
> Mississippi Article 15, Section 265: No person who denies the existence of a Supreme Being shall hold any office in this State.
>
> North Carolina Article VI, Section 8: The following persons shall be disqualified for office: First, any person who shall deny the being of Almighty God.
>
> South Carolina Article XVII, Section 4: No person who denies the existence of a Supreme Being shall hold any office under this Constitution.
>
> Tennessee Article IX, Section 2: No person who denies the being of God, or a future state of rewards and punishments, shall hold any office in the civil department of this state.
>
> Texas Article 1, Section 4: No religious test shall ever be required as a qualification to any office, or public trust, in this State; nor shall any one be excluded from holding office on account of his religious sentiments, provided he acknowledge the existence of a Supreme Being.

Creationism takes many forms, each concocting its own set of "testable predictions." To name a few: all organisms were designed in their final form, so no intermediate forms should appear in the fossil record; radioactive decay "sped up" during the Biblical flood, yielding the impression the earth is older than it really is; humans and dinosaurs existed simultaneously, their fossils somehow deposited in separate and discrete strata during the global inundation; biology expresses elements of "irreducible complexity", such as the eye, that could not have arisen from evolution; "junk DNA" should be absent in all life forms, as their traits were perfectly designed for their habitats. When the predictions are disproved, as all of the preceding have been, creationists simply fabricate new sets of predictions in an endless game of musical chairs.

The election between *adopt* and *deny* as a means to alleviate cognitive dissonance is not as straightforward or binary as presented. The election is driven by the interaction among cognitions - the assorted perceptions formed by our mental processes. The formation, dissolution, relevance, magnitude, and relationship among cognitions guide our thoughts and behaviors.

Selected psychological, sociological, neuroscientific, and evolutionary explanations of religious belief.

Diagram 8.1

Explanations

1. Religion as Psychological Palliative
2. Religion as All-Encompassing Explanation
3. Religion as Mental Release
4. Religion as Cultural Construct
5. Religion as Paleolithic Adaptation
6. Religion as Neurochemical Reaction
7. Religion as Evolutionary By-Product

[IX]

Perspectives

What can be asserted without evidence can also be dismissed without evidence.[302]

CHRISTOPHER HITCHENS (1949–2011) ANGLO-AMERICAN AUTHOR

From the respective *what to think* bodies of knowledge from Religion and Reason, let's look at how each domain responds to some of life's thorniest questions.

IS THERE A SUPREME BEING?

Religion responds to this question unequivocally. Yes, there is a supreme being, or multiple ones, depending on belief system. Such an assertion, on its own, might suggest a tautology. Indeed, self-referential and true-in-all-instances declarations of a divine presence crop up in all sacred texts. As the Bible attests: "I am God Almighty."[303] Yet Religion offers other arguments for the existence of a deity, the most common of which fall into three categories: teleological, cosmological, and ontological.[304]

The origin of the teleological argument, contemporarily branded as creationism or intelligent design, is unclear, but dates at least to the ancient Greeks. The argument contends that our world exhibits endless signs of complexity, elegance, and purpose: the intricacy of a living cell, the serenity of a pristine alpine lake, the sweet birdsong of a nightingale. The line of reasoning then extrapolates from specific cases to general themes: laws of nature perfectly attuned for our existence; an intricacy of detail underlying all life forms; innumerable species perfectly suited to their environments; the awe-inspiring, limitless beauty of the entire cosmos. The path from general themes to obvious conclusion is direct: Intelligent design, ergo an intelligent designer as creator of all.

The genesis of the cosmological argument also dates to the ancient Greeks, possibly Aristotle. The argument begins by asserting that the mere fact that anything exists, rather than nothing, requires an explanation. The explanation derives from causation, that everything has a cause. Before *H. sapiens* can exist, the earth must exist. Before the earth can exist, the universe must exist. Before the universe can exist, the Big Bang must occur. Tracing backward in this manner from cause to cause, precluding infinite regress, leads to a first cause. The first cause, the cause of the Big Bang and all creation, must be a supreme being.

The Italian monk and theologian, Anselm of Canterbury (1033-1109), is credited as the originator of the third argument, the ontological argument. The ontological argument makes no comment regarding the universe or its contents, but affirms a connection between ideas and underlying reality. Because the idea of a deity can exist in the mind, it can likewise exist in reality. An understanding of God as the apotheosis of the most perfect idea thus bears witness to his existence in both the spiritual and temporal worlds.

The teleological, cosmological, and ontological arguments have been scrutinized and debated for centuries, both within and outside Religion. Unsurprisingly, those who already accept the existence of a deity embrace the arguments. Equally unsurprising, Reason rejects the arguments. A full treatment of the logical fallacies inherent in the arguments is beyond the sweep of this book, but the following criticisms provide a sense of the debate:

TELEOLOGICAL The teleological argument follows the familiar practice of Religion by offering no convincing evidence, no testable predictions, and no independently verifiable conclusions. It disregards alternative explanations of observable, natural phenomena. It dismisses abduction, the selection of the hypothesis that best fits the facts. It also ignores myriad ancillary questions. If robust evidence signaled deliberate design, what of the designer? If the designer is a deity, which one? If the designer is a member of an advanced civilization, from where? And, to repeat a theme, if testable predictions were offered and falsified, Reason contends that Religion would not accept the unwelcome conclusion.

COSMOLOGICAL The cosmological argument is self-refuting. It argues for a deity as first cause, but does not challenge the cause of the deity, ignoring its basic tenet that everything has a cause. It simply asserts that its deity is eternal and has perpetually existed. There is no explanation of the property that allows a deity to exist without cause, yet somehow precludes a universe from existing without divine cause.

ONTOLOGICAL The ontological argument is perhaps the most widely refuted. The English philosopher David Hume (1711-1776)[305] and the German philosopher Immanuel Kant (1724-1804),[306] among others, articulated tenable refutations. One of the earliest rebuttals emanates from the core premise of the argument. If an idea implies a reality, then other "most perfect" ideas must perforce exist: the myriad deities avowed by other religions, secular utopias, and similar chimeras. Further, the ontological argument is non-falsifiable, ergo nugatory.

This is an opportune moment to reiterate a point from *The Case of the Earth's Age*. Argument occupies a distinguished billet in Religion. Theologians employ argument to substantiate their claims, debate doctrinal differences, and influence adherents. Unencumbered by evidence, they parse sacred texts, sentence by sentence, word by word, even letter by letter, devising scriptural interpretations to bolster their contentions. Others rely on protestations of divine guidance to underpin their pronouncements. With millennia of experience, Religion is adept at crafting superficially appealing arguments, each anchored by theistic authority and dispensed by charismatic personages. And yet. Reason recognizes that the best-reasoned argument, unsupported by empirical evidence and testable predictions, is meaningless. The intelligence, credentials, reputation, and authority of the proponent are equally pointless. They serve no purpose in avouching an argument's validity.

There is a method, of course, by which Religion can uphold its arguments and effectively demonstrate the existence of a supreme being. Provide evidence consistent with the scientific method that a deity exists. Provide evidence their deity, and only their deity, created the universe. Provide evidence that explains why their deity created the universe some 13 billion years ago, not earlier or later. Provide evidence their deity is a loving, caring deity (where such qualities are proclaimed), rather than a remote and indifferent deity. Provide evidence their deity is relevant, while the thousands of deities from other belief systems are irrelevant. And provide evidence their deity is indeed a deity, not merely an advanced life form that mimics the prowess of a deity.

Has Religion, over the millennia of its existence, furnished any such evidence? None. Absolutely none. This reality leads Reason to dismiss faith-based arguments for a supreme being as nothing more that wishful thinking. Reason likewise challenges the superfluous nature of deities. In disciplines where Reason is able to explain a phenomenon, such as the complexity and diversity of life on Earth, appending a deity offers no additional explanatory power. And, equally, in spheres where Reason cannot yet provide an explanation, invoking a godhead yields no meaningful insights. Even worse, it often serves only to terminate further inquiry.

Let's close this section with a thought experiment, one that, once again, discloses Religion's aversion to testable predictions. The faithful often believe in the power of prayer, that their deity listens to prayers and sometimes intervenes in the world to grant petitions. Envision a randomized, double-blind experiment where one group of people prayed for a predefined outcome, say, fewer injuries caused by automobile collisions in a well-defined geography. A separate contingent could serve as the control group, uttering no prayers about vehicle accidents in a separate geography. Repeat the experiment scores of times around the world, assessing the consequences by measuring auto casualties in the target geographies. A statistically significant decline in casualties could provide evidence of Religion's assertions regarding the power of prayer. But if the complexities of such an experiment could be solved, would Religion agree to conduct the trial? If the results showed no meaningful decline in casualties, would Religion accept the meaninglessness of prayer? The response to both questions is no. The risk to Religion is too high, as a null result would further verify the already well-documented placebo effect of prayer, reaffirming its role as wishful thinking with no consequence on the external world.[307]

WHAT IS THE ORIGIN OF THE UNIVERSE?

Again, Religion's response is categorical: "Behind the world stands its author and sovereign ruler, God."[308] In Christianity the Bible declares, "In the beginning God created the heaven and the earth."[309] In Islām the Qur'ān states, "Your Lord is Allāh, who created the heavens and earth in six days and then established Himself above the throne."[310] Sikhism recognizes, "[The] One Universal Creator God."[311]

Creation stories vary, but all share singular traits. We mentioned two in *The Case of Human Origins*: life originates where believers of the myth happen to live, and the tales assert the existence of one or more supernatural beings. Now let's mention two more. By definition, each belief system attributes creation to its deity and to no other. What's more, the creation yarns provide no detailed explanation of how the creation was accomplished. They offer only vague and picayune assertions, each shrewdly packaged into fables that contradict all known physical processes.

Reason, in contrast, bases its response on a theory named the Big Bang, a brief version of which goes something like this: Approximately 13 billion years ago a singularity, an infinitely dense and high-temperature point containing all the mass and energy of the universe, many times smaller than an atom, began to expand. Within the first second, the four fundamental forces of nature (the strong force, weak force, electromagnetic force, and gravity) emerged, and the first elementary particles (quarks) formed. Photons and atomic nuclei emerged ten seconds after the Big Bang and, along with other sub-atomic particles, dominated the universe. After 380,000 years, as the universe expanded and cooled, the first atoms (hydrogen) began to form. The same atoms, 300 million years later, coalesced into the first planets, stars, and galaxies. Nuclear processes within the stars forged heavier elements, such as carbon. When the most massive of the stellar objects reached the end of their lives, they exploded (as supernovae), spreading their matter into the expanse of space, to be incorporated into the next generations of stars and planets, such as Earth. It is a fantastic story, pieced together gradually and meticulously over the last 100 years. Many scientists contributed to the theory, but let's review three of the discoveries that perhaps best illustrate the journey to the theory.

In 1905 the German physicist Albert Einstein (1879-1955) published his special theory of relativity. The theory provided us with the now famous equation, $e=mc^2$, which illuminates the relationship between energy and mass. $E=mc^2$ explains, among other things, the fusion process that occurs in the cores of stars, converting mass to energy, and creating many of the elements necessary for life, including carbon. Ten years later, Einstein published his theory of general relativity, which combined space and time into a single property, spacetime, that curves, contracts, and expands in the presence of mass. Einstein showed that spacetime is a property of the universe, indicating that space and time did not exist prior to the Big Bang.

In the 1920s the American astronomer Edwin Hubble (1889-1953), from observations recorded at the Mount Wilson Observatory in California, produced two notable discoveries. First, Hubble demonstrated that stars and galaxies exist outside our own Milky Way galaxy, indicating the universe was considerably larger than previously thought. Until then, accepted thinking placed all stars within our galaxy. Second, by measuring the redshift of light from stars and galaxies, Hubble confirmed that these objects were receding from the Milky Way. The more distant the object, the faster it was receding. This provided the first observational evidence the universe was expanding. Hubble's findings confirmed the theories of other physicists, such as the Belgian priest Georges Lemaître (1894-1966), who also predicted the expansion of the universe. The confirmation of an extensive and expanding universe implied that, if you could reverse the clock, stars and galaxies would move closer together, ultimately merging to a common point of origin. The rate of recession thus furnishes a means of estimating the age of the universe. This technique indicates the universe is roughly 13.8 billion years old.

In the 1940s the American physicists Robert Alpher (1921-2007) and Robert Herman (1914-1997) suggested that thermal radiation from the early universe, originating 380,000 years ago and known as Cosmic Background Radiation (CMB), could be detected. They predicted its temperature (5° Kelvin) and its location in the microwave region of the electromagnetic spectrum. In 1964 the American physicists Arno Penzias (1933-) and Robert Wilson (1936-) discovered the CMB using a radio antenna at Bell Laboratories in New Jersey. The discovery confirmed the high temperatures and pressures of the nascent universe, a critical prediction of the Big Bang theory.

As implied by the preceding discoveries, the power of the Big Bang theory resides in its ability to explain a broad set of phenomena, from the origin of the universe to its present-day structure, informed by deep and extensive experimental evidence. What's more, the theory dispelled competing proposals, notably the Steady State theory, which argued that new matter was constantly being formed to fill the gaps in an expanding universe. Yet despite the success of the Big Bang model, open questions remain. What caused the Big Bang to bang? What happened one second prior to the Big Bang? Was the Big Bang the beginning of everything, or simply a transition from some prior state? What is dark matter? What is dark energy? How does quantum mechanics give rise to the reality we perceive? How can quantum mechanics be reconciled with general relativity? What explains the values of the fundamental constants of the universe? How large is the entire universe relative to the observable universe? Do additional dimensions exist beyond the four of space-time? New discoveries that answer these questions might compel adjustments to the Big Bang theory or, as always with Reason, drive the emergence of a new theory.

Let's spend a moment longer on the concept of singularities. A singularity is difficult to imagine, even more difficult to explain. It is suggested, but not proven, that energy fluctuations caused the singularity to expand, creating the universe we inhabit. Some physicists argue that if there was one singularity, there might be many, each expanding to create its own universe, each universe possessing its own space, time, matter, energy, and - possibly - its own physical laws. This assortment of universes, each arising from its own Big Bang, is tagged a multiverse. Within the multiverse, perhaps we just happen to reside in a universe whose physical laws enable our existence. At present, there is limited evidence for this idea. Many years likely will pass before advances, both in theory and technology, permit testing of the conjecture.

A further conjecture suggests the universe as a physical reality is an illusion, its inhabitants all residing in a digital simulation. The response to such speculations, outlandish as they initially might appear, is not to dismiss them. In Reason, the response is to repeat the words of Sagan: Extraordinary claims require extraordinary evidence.

WHAT IS THE ORIGIN OF LIFE?

This is another query Reason cannot yet answer. But Reason has not remained idle. In 1952 American chemists Stanley Miller (1930-2007) and Harold Urey (1893-1981) conducted an intriguing experiment.[312] They decided to re-create the conditions of primeval Earth. To represent the atmosphere they placed ammonia, hydrogen, and methane (gases common in the early history of the earth) in an upper flask. To represent the ocean, they placed water in a lower flask.

They then connected the two flasks with rubber tubing to form a sealed circuit, such that heating the lower flask would induce vapor to rise into the upper flask, and cooling the upper flask would cause liquid to precipitate into the lower flask. Lastly, they inserted electrodes in the upper flask, firing electrical sparks to simulate lightning in the atmosphere. Then they waited. Upon examining the flasks one week later, they identified that a number of amino acids - the building blocks of proteins and a basic ingredient of life - had formed in the liquid.

The Miller-Urey experiment did not give rise to life, but it demonstrated that organic compounds intrinsic to life arise naturally. As yet, however, there is no accepted theory on how the first compounds formed into self-replicating molecular systems, then into the first primitive biological life. But Reason is exploring distinct avenues: the possibility of microbial life originating below the surface of the earth; of lipids (fats, waxes) coalescing into the first self-replicating life forms; of fast tidal cycles concentrating molecules into pools rich in protonucleic acids;[313] of specific clays acting as catalysts in molecular bonding; of complex molecules arriving on Earth from space (comets, asteroids, interstellar dust); of life arising in iron- and sulphide-rich hydrothermal vents on the seabed; and of the self-sustaining decay of radioactive elements providing the energy to hammer out life's building blocks. Still another possibility is panspermia, the idea that life itself (not merely its molecular building blocks) arrived from asteroids, meteoroids, comets, or other bodies in the Solar System. Note that panspermia, if confirmed, would not explain the origin of life, but merely shift the question of its origin.

In addition to the still-unknown chemical and biological processes that led to the creation of life, a more rudimentary question regards the nature of the universe. Somewhere between five and forty constants (the number varies by source) possess values that appear "just right" for the emergence of life, leading some to argue that the universe appears arbitrarily fine-tuned for life. An example of one such constant is the strong force, the force that holds protons and neutrons together in the nucleus of atoms. If the strong force were ever so slightly more powerful, no hydrogen would have remained from the Big Bang. No hydrogen means no water (a hydrogen-based compound) and no fuel for stars. If the strong force were slightly weaker, no elements heavier than hydrogen could have arisen. No elements equals no life.[314] Other "just right" constants include the decay rate of protons, the mass of neutrons relative to the mass of protons, and the strength of the force of gravity relative to the strength of the electromagnetic force. Reason can measure the value of these constants but is not yet able to explain why each constant possesses its respective specific value.

I discuss fine-tuning of the universe, as it opens a door Religion (and creationists in particular) cannot resist entering. Creationists cite the five-to-forty parameters as evidence of the fine-tuning of the universe, then eagerly fault Reason for its inability to explain the ostensible fine-tuning. Reason's lack of explanation allows Religion to claim its deity designed the universe specifically to create life. Creationists claim, for example, that a deity set the force of gravity just strong enough so that galaxies, stars, and planets could form, but not too strong; otherwise all matter would collapse together. Religion hurriedly adopts this premise as ratification of its beliefs.

Yet the creationist argument is an empty theory, in that it offers no detailed explanation for its claims. What is the range of acceptable values for each of the 40-odd parameters that permit life? Why does an electron weigh some 1800 times less than a proton?[315] Why is the mass of proton 99.86% that of a neutron? Why is gravity some 1034 times weaker than the electromagnetic force? How do biological processes first emerge from an abiotic environment? Why do we breathe oxygen, not carbon-dioxide as plants do? Why do we live in a discrete, quantum universe of energy, not in a continuous, analog universe? Why is the universe chiefly a lethal vacuum? Religion cannot supply answers to these and similar questions, claiming only "if Reason cannot explain it, our chimerical version must be correct" - yet another instance of the *God of the Gaps* argument.

Unsurprisingly, Religion's ingrained response to the basic question in this section is clear. A deity created all life. Period. The deity's method of creation varies by religion, as does its motivation, which might stem from love, spite, or indifference. Or creation might serve solely as a test bed for the possible salvation of selected members of humanity.

IS LIFE STILL EVOLVING?

Religion offers two opposing responses to this question. The first pertains to belief systems that now accept evolution by natural selection. All such faiths respond with a cautious yes. To safeguard their dogmas, however, they contrive an accommodation. They accredit to their deities the role of master designer; here, a divine author who devised and initiated the process of evolution. This interpretation, in the eyes of the faithful, aligns theistic doctrine to modern scientific findings.

Religion's second response is thornier. A faith that posits its deity as the creator of all life in its final and unchanging form must, by definition, disavow evolution. It must declare life is not evolving. In this case, the faith is forced to adopt one of the aforementioned techniques to combat scientific evidence: exploiting gaps in scientific understanding, altering the meaning of evidence, discrediting empirical findings, or dismissing confirmed theories. The faith must, in short, banish science to retain its myths.

Reason responds to this inquiry with a categorical yes. Changes in environments continue to drive adaptations and changes in species. The strongest example of evolution in *H. sapiens* over the last 5,000 years is resistance to infectious diseases. The transition from hunter-gatherer to agricultural societies caused humans to live in closer, constant proximity to each other and to domesticated animals.[316] This change in environments facilitated the transmission of infectious diseases, with individuals responding to the spread in one of two ways. Those with natural resistance passed the trait to their offspring, slowly propagating immunity among the general population. Sadly, those without resistance often perished. Native Americans, living apart from Europeans until the fifteenth century, evoke a tragic example. Unexposed and susceptible to European contagions, 70% to 95% of Native Americans succumbed upon contact with European settlers.

The Native American tragedy demonstrates that evolution does not universally equate to progress, at least not for *H. sapiens*, and at least not in the short term. Contrary to common perception, evolution also does not necessarily drive complexity. Natural selection sometimes produces anatomically simpler, less complex species, such as organisms who surrender their eyesight or their pigment by dwelling continuously in dark habitats. The salient point is that natural selection produces organisms that are "good enough" to survive in their environments, rather than perfectly designed species.

Evolution offers another pertinent insight. Religion posits a model where a higher intelligence creates life forms of lesser intelligence. Darwin inverted the model, demonstrating that natural selection is a non-intelligent process that creates life forms of higher intelligence. The culmination of this process (so far) is the self-aware, contemplative, intelligent species named *H. sapiens*. Our species is the intelligent designer, the organism created by evolution, that now sits on the cusp of extending both natural and artificial selection to new heights. Advancements in knowledge and technology permit us to alter our own genes and those of other species. GMO crops and bacteria engineered to manufacture insulin are but early examples. Coupled with the rise of biotechnology, artificial intelligence, and boundless data, perhaps we are on the verge of harnessing our own evolution and our own destiny.

ARE WE ALONE IN THE UNIVERSE?

Here again, Religion's response varies. Ancient scriptures typically dispense no comment on extraterrestrial life, as the topic was of no consequence when the religions first arose. It is accordingly unclear how older religions would react to confirmation of life on a distant planet. The dilemma might be more straightforward for selected

Asian religions, such as Hinduism and Buddhism, which already hold varied cosmological beliefs. And perhaps simpler still for religions established in the nineteenth century: as knowledge of the cosmos expanded, these newly formed religions often contemplated the idea of non-terrestrial life. Yet all faiths, both ancient and modern, will have to fine-tune at least some elements of their beliefs. Consider some hypotheticals.[317]

- Judaism rejects a claim from an alien life form, one who possesses no knowledge of Abraham or Moses, yet declares himself a Jew.
- Catholics conclude far-off organisms are subject to original sin, thus requiring redemption.
- Lutherans decide the aliens cannot be Christians, as the incarnation of Jesus Christ occurred on Earth.
- Orthodox Christians classify the cosmic interlopers as fallen angels, acting on the orders of Satan.
- Evangelicals, noting man was created in God's image, denounce the intelligent aliens, who float through the atmosphere of their planet as multi-eyed, hairless orbs.
- Christian fundamentalists, who interpret the Bible literally and believe God's sole intent was to create humanity on Earth and nowhere else, anxiously await new revelations from the Lord.
- Southern Baptists espy an opportunity for missionary work.
- Anglicans craft an exotheology that extends God's grace to their distant brethren.
- Mormons celebrate the truth of their scriptures, which foresaw the discovery of alien life forms: "That by him, and through him, and of him, the worlds are and were created, and the inhabitants thereof are begotten sons and daughters unto God."[318]
- Hindus, Buddhists, and members of Bahá'í Faith also rejoice, as their faiths encompass all beings in the universe.
- Sikhs similarly embrace life elsewhere, but find the aliens befuddling. It is unclear how Sikhism's Five Articles of Faith - which place religious significance on hair, combs, bracelets,

swords, and undergarments - apply to the aliens, whose bodies contain no appendages.
- Sunni scholars require the far-flung creatures to pray five times per Earth day, facing Mecca as they do so.
- Shia theologians similarly demand that the life forms, at least once in their centuries-long lives, undertake a galactic pilgrimage to Mecca.
- Jainists pity the aliens, as life in human form offers the only chance to escape samsāra, the endless cycle of rebirths, each one imbued with suffering and misery.

The above might appear tongue-in-cheek, but Religion has hotly debated less consequential questions for millennia. When intelligent life is discovered on a distant sphere, new disputes will arise, and new schisms will develop. Theologians will formulate their arguments, each buttressed from scriptural decipherment or timely revelation. Prophets, clerics, and laypersons will vie to be the next Aquinas, performing a similar thirteenth-century reconciliation between the teachings of Aristotle and Christianity, this time re-interpreting sacred texts to subsume the reality of extraterrestrial life.

Reason cannot yet determine whether we are alone, as the capability to detect extraterrestrial life only emerged over the last fifty or so years. One detection technique is to employ immense radio telescopes, such as the 305-meter radio telescope at the Arecibo Observatory (now inoperative) in Puerto Rico or the 500-meter Aperture Spherical Telescope in China, to detect signals from other civilizations. Another is to use spectroscopy to identify the chemical fingerprints of life in the atmospheres of other worlds. Yet another, perhaps more likely possibility, is to detect traces of life on nearby objects with earth-launched probes. The probes often seek out liquid water, which is considered a prerequisite for life. To date, water has been confirmed on several nearby objects: the planet Mars, the asteroid Ceres, Saturn's moon Enceladus, and Jupiter's moons Europa and Ganymede. It is uncertain, however, when these or other techniques will detect signs of life or intelligence.

Nonetheless, Reason can apply a probability to the uncertainty. One method is based on the Drake equation, which breaks the query into manageable parts.[319] The American astrophysicist Frank Drake (1930-) formulated the equation in 1961 as $N = a * b * c * d * e * f * g$, where each letter equates to a component in Diagram 9.1.

Observations based on current technology provide values for the first three components. Estimates for the remaining four vary more widely. In 2011 the Astronomical Society of the Pacific (ASP) provided its estimates for all seven components of the Drake equation.[320] Their numbers, which appear in parentheses after each component, suggest $N = 500$. In other words, there might be 500 communicating civilizations in our galaxy. And remember, that's just our galaxy.

The number for the entire universe would be $N * h$, where h denotes the number of galaxies. What might N equal for the entire universe? Let's be conservative and first reduce N, as calculated by the ASP, by a factor of 10 million, such that $N = 0.00005$, far less than one communicating civilization per galaxy. Now, how many galaxies are in the universe? Based on observations from the Hubble Space Telescope and other sources, the Israeli astrophysicist Mario Livio (1945-) calculates there are hundreds of billions of galaxies in the universe. Again, let's remain conservative and select a number on the low end, say, $h = 100$ billion. This suggests there could be 5 million (N times h) communicating civilizations in the universe. Of course, our ability to detect a signal from a far-off alien world is low.

Diagram 9.1
The components of the Drake equation.

- (N) The number of civilizations with whom communication might be possible
- (a) The average rate of star formation in our galaxy (1.0 per year)
- (b) The fraction of formed stars that have planets (0.4)
- (c) The average number of planets per star that can support life (1.0)
- (d) The fraction of planets from (c) that actually develop life (0.5)
- (e) The fraction of planets from (d) that develop intelligent civilizations (0.5)
- (f) The fraction of civilizations from (e) that develop communications (0.5)
- (g) The length of time over which the civilizations from (f) release detectable signals (10,000 years)

(N) $= a * b * c * d * e * f * g = 500$ communicating civilizations

For practical or emotional reasons, you might find these numbers implausible. Two civilizations attempting to communicate with one another might exist millions of light-years apart in distance and millions of years apart in time. So let's apply the Drake equation another way, calculated only from its first four components. First, let's reduce the ASP's estimates for these components to 0.1 each. And second, let's lower our already conservative estimate for *h*, the number of galaxies in the universe, to 10 billion. This give us $N = a * b * c * d * h$. The solution to this modification of the Drake equation, as depicted in Diagram 9.2, suggests that one million planets harbor some form of intelligent or non-intelligent life - a much reduced but not inconsequential number.

Admittedly, the Drake equation can yield a wide range of outcomes. So, too, do variants of the equation, of which there are many. Some base their calculations on the biosignature of gases. Others consider colonization rates: the likelihood that once life emerges, it spreads to other planets. Still others take into account reappearance rates: the probability that an intelligent species arises then disappears, to be followed by the emergence of a different intelligent species millions of years later on the same planet. Regardless of approach, one aspect is certain. The value of N is greater than zero. Why? Because we know intelligent life arose at least once - on planet Earth.

To many, it is inconceivable that life emerged only one time in the 13-billion year history of the universe, never to arise elsewhere. Reason, therefore, will continue its search for extraterrestrial life. When the day of its discovery arrives, a third displacement will occur. Recall that Copernicus initiated the first dislocation, expelling our species from the center of the cosmos. Darwin set in motion the second, dispelling the notion we are a favored, purposely-designed species. When life is detected on a distant world, we will become but one of countless life forms scattered across the universe, though still precious on our own world.

Diagram 9.2
A variation of the Drake equation estimating the number of planets that harbor life.

(N') The number of planets supporting life in the universe
(a) The average rate of star formation in our galaxy (0.1 per year)
(b) The fraction of formed stars that have planets (0.1)
(c) The average number of planets per star that can support life (0.1)
(d) The fraction of planets from (c) that actually develop life (0.1)
(h) The number of galaxies in the universe (10 billion)

(N') = a * b * c * d * h = 1 million planets

ARE MORALS ABSOLUTE?

Religion's response to this query yet again is explicit: Morals are absolute. The five sacreds articulate and reinforce the moral principles of each faith. Whether explicitly stated in a sacred text, such as Christianity's Ten Commandments, or interpreted by sacred persons, each faith exhorts its disciples to abide by its moral philosophy. Each philosophy, in turn, provides guidance on a broad range of topics, including marriage, divorce, free will, charity, intoxication, slavery, adultery, stealing, abortion, homosexuality, humbleness, honesty, frugality, tolerance, love, and respect. Religion, in essence, provides a handbook on how to lead a moral life.

Reason provides a more nuanced response, one that offers insights into the origin and nature of morality, and one that requires a fresh visit to Zaug and his kinfolk. Zaug's tribe, comprising a few score denizens, formed part of a larger clan of several hundred individuals, a typical societal model in the hunter-gatherer period. Members of the clan shared common attributes: a common language, a common ethnicity, a common belief system, and a communal use of land. The defining characteristic of clans, however, was kinship - most individuals were at least distantly related to each other.

The relatively small size of kinship societies and the inter-relatedness of their members fostered bonds of interdependence and trust. Conflict, when it arose, was settled by the informal precepts of the clan. Such precepts formed the basis for in-group behaviors, the expected behaviors for members of the clan. The establishment of in-group behaviors, by definition, implies the formation of out-group behaviors: how members of the in-group react to and treat out-group individuals who are not members of the clan. Zaug's clan often regarded out-group individuals as a threat, to be approached cautiously or expelled from the area. Thus, in-group behaviors served as a protection mechanism for the clan.

During the transition from hunter-gatherer to agricultural societies, clans merged into what Diamond refers to as chiefdoms, and ultimately into states.[321] Chiefdoms and early states comprised thousands of individuals. Within these larger, multi-ethnic, multi-language, and multi-faith societies of unrelated individuals, the informal rules of kinship no longer sufficed.[322] Furthermore, as we saw in *The Origin of Religion*, the rise of agrarian societies introduced new conventions and methods of organization, such as division of labor, private property, stratified social structures, and centralized institutions of authority. In consequence, these societies required more robust moral philosophies to manage conflict, build trust, and foster cohesiveness. It therfore comes as no surprise that belief systems emerging during this period encompassed extensive moral and ethical philosophies, as the societies themselves acted as a catalyst for their creation and adoption.

The preceding narrative provides the foundation for Reason's answer to the question of absolute or relative morals. Morals are relative. Moral philosophies developed in parallel with the evolution of human society, becoming more detailed and complex as societies themselves grew more complex. As such, moral frameworks arose not from eternal sacred laws existing outside of human experience, but emerged at specific moments in history, shaped by the minds of humans as the need arose. Religion, of course, rejects this view, choosing instead to maintain the mythos that ethical codes emanate from ethereal sources. Religion thus accords to itself a self-manufactured authority not only to ordain a litany of admissible and proscribed acts, but to reward or punish all who fall under its grip.

Yet Reason argues that Religion is not a prerequisite for positive moral behavior. Societies can, and do, develop equally effective moral philosophies without appeal to Religion. Thomas Hobbes (1588-1679), John Locke (1632-1704), Jean-Jacques Rousseau (1712-1778), and Jeremy Benthem (1748-1832) are four of the philosophers who provided a scholarly, reasoned basis for a social contract, a type of societal agreement encompassing the authority of rulers, the rights of the governed, the education of citizens, and the organizing rules of society. Reason contends that secular education, both formal and informal, capably imparts morals. Leaders can, and do, reinforce nonreligious moral principles through both word and deed. So, too, do community traditions and publicized exploits of notable, desired behaviors. The outcome is no societal anomie, no dystopia of moral chaos. Secular societies, built on a shared humanity, develop moral philosophies that are indistinguishable from Religion: be loyal rather than disloyal, help rather than harm, understand rather than judge, love rather than hate. Individuals in societies that subscribe to these principles live equally valid moral lives. The secular individual, as such, is every bit as moral and principled as the pious - often more so. And interestingly, countries whose inhabitants express low levels of religious belief - Denmark and Sweden, for example - often score highest on international measures of social equality, societal stability, and happiness.[323]

Unsurprisingly, Religion discounts relative values in favor of a strict moral code based on absolute rights and wrongs, a code underpinned by heavenly authority. To the steadfast advocates of absolutes, however, it is a short mental step to a next, all-too-common enterprise: imposing their absolutes on everyone. Witness the restrictive treatment of secular bloggers in present-day Bahrain, Bangladesh, Indonesia, Iran, Malaysia, Pakistan, Saudi Arabia, Singapore, and Turkey. Reflect on the countries where religious education is required, with no secular alternative: Algeria, Iraq, Morocco, Somalia, United Arab Emirates, and Zimbabwe. Contemplate the nations where criticism of religion is flatly outlawed: Afghanistan, Egypt, Guyana, Kazakhstan, Nigeria, Sudan, Suriname, and Tanzania.[324] The enumerated countries, plus many more, affirm a positive correlation between divine absolutes and legislated intolerance.

The aforementioned Rousseau is a further case in point, albeit historical. His writings on politics (arguing that power resides with the people, not divinely-empowered monarchs), education (proposing the teaching of reason to children), and religion (suggesting all religions are equally worthy) inflamed French and Swiss ecclesiastical authorities, who accused Rousseau of heresies. The same persecutors, blindly bound to their absolutes, ordered Rousseau's books banned and burned, issued warrants for his arrest, evicted him from his home, hounded him across Europe, and forced him to endure life as a fugitive. From his ill-treatment, Rousseau's view on Reason is worth reciting: "The highest accomplishment of education is to make a reasoning man ... If children understood how to reason, they would not need to be educated."[325]

DOES RELIGION PROMOTE VIOLENCE?

Athanasius the Apostolic (c.296-373), who served as the bishop of Alexandria for 45 years, was a proponent of the Christian doctrine that the Son (Jesus Christ) and the Father (God) are identical in essence - homoousios. Arius (c.250-336), a Christian priest in Alexandria, championed a competing doctrine, that the Son and Father bear similarities, but are not identical - homoiousios. The contrast in belief provoked mob riots and deaths among the followers of each doctrine, ultimately forcing the Roman Emperor Constantine to call the first ecumenical council in 325 CE to resolve the quarrel. Christian bishops attending the First Council of Nicaea decided in favor of homoousionism. Years later, the English historian Edward Gibbon (1737-1794) commented on the Christian brutality, noting it arose from the battle over a single diphthong - hom***oou***sios versus hom***oiou***sios.[326] You might even say over a single letter, the letter *i* - homoousios versus homo***i***ousios.

To an individual with no vested interest in the outcome, a barbed exchange over a diphthong seems trivial, even non-sensical. Yet theological discord, when perceived as a challenge to core beliefs, invokes strong passions in Religion, leading to the accusation Religion promotes violence. The accusation commonly rests on three arguments: scripture, monotheism, and group dynamics. Let's examine each in turn.

The scriptural argument, as the name implies, derives from passages in sacred texts, passages which explicitly advocate violence. Consider three instances from the Bible. The first sanctions not only bloodshed but the treatment of women as spoils of war: "Now therefore kill every male among the little ones, and kill every woman that hath known man by lying with him. But all the women children, that have not known a man by lying with him, keep alive for yourselves."[327] The second instance licenses stoning to death a bride who, on her wedding night, was not a virgin.[328] In the third, the deity slaughters 50,070 males for having "looked into the ark of the Lord.[329] Similar examples abound.

A common reaction from Religion, while acknowledging the truth of scripture, is to downplay the atrocities as requisite in their context but no longer pertinent. Continuing the example from the Bible, defenders contend that the justice-based deity of the Old Testament is succeeded by the loving and tolerant deity of New Testament. Yet the defense is specious, as it offers no plausible answers to myriad related questions. What prompted the deity to initiate a lengthy progression of brutal and merciless acts? Why were the acts necessary only during a select interval of human history? Why were the acts confined to specific peoples in a restricted geography? What led the truculent deity to cease hostilities in favor of a "loving and tolerant" successor? How benevolent and fair-minded was the successor? Consider Christ's fiat in Luke, directed at those who abstained from his authority: "But those mine enemies, which would not that I should reign over them, bring hither, and slay them before me."[330]

A related defense is to declare that violence stems from misinterpretation and misuse of Religion, rather than from Religion itself, in essence blaming rogue operators. Pope John Paul II (1920-2005) employed this counter-argument in his *Symposium on the Roots of Anti-Judaism* in October 1997. Christ's Vicar on Earth acknowledged anti-semitism, reaffirming it "has no justification and is absolutely reprehensible." As to cause, he was careful to distance the church by stating, "I do not say on part of the Church as such."[331] Instead, he attributed the discriminatory treatment to erroneous interpretations of the New Testament, articulated by philistine disciples of Christ.

The Qur'ān provides its own examples of what we today deem crimes against humanity: "Kill the polytheists wherever you find them and capture them and besiege them and sit in wait for them at every place of ambush."[332] Witness also, "Those who disbelieved will have cut for them garments of fire. Poured on their heads will be scalding water."[333] Similar extracts from the Qur'ān and the

Hadīth condone forced conversion,[334] ethnic cleansing,[335] the killing of apostates,[336] and the beating[337] and stoning[338] of women.

Argument from scripture often is a fool's errand, as context and interpretation vary widely. Suffice it to say, however, that the Bible, the Qur'ān, and other sacred texts offer numerous comparable citations, providing instructions on genocide, rape, rapine, and the destruction of families. Indeed, the texts frequently present wanton acts as morally obligatory. Myriad popes, preachers, imāms, clerics, and *soi-disant* evangelists leverage the scriptural declarations to propagate near endless violence. The French abbot, Bernard of Clairvaux (1090 - 1153), evinces the practice. To obtain new Christian lands, he promulgated a "convert or exterminate" philosophy towards pagans and other non-believers, one of many such examples from nearly every major belief system in history.[339]

The second argument for violence in Religion stems from the rise of monotheism. This argument suggests that belief in a single deity breeds divisiveness, inducing an absolutist mentality of "my deity is valid and yours is false." The argument extends to polities that embrace monotheistic religions. Proponents of this argument base their claim, in part, on comparative history, cataloging and analyzing the nature of conflicts. The Crusades, for example, championed the "truth" of Christianity over the "heresy" of Islām. Present-day Islāmic fundamentalism reverses the claim, asserting the verity of Islām over the blasphemy of Christianity.

The same proponents research religious texts, examining the number and severity of violent acts toward non-believers. Monotheist gospels, they conclude, eclipse their polytheistic competitors in admonishments to destroy infidels, smite apostates, and cast unbelievers into the bowels of hell, relegating them to spend an eternity in the infernos of the netherworld. The scriptures of Buddhism and Hinduism, by contrast, express fewer absolutes, acknowledging multifold paths to spiritual fulfillment. Proponents thus contend that polytheist belief systems urge greater tolerance of sundry deities and their legions of followers - at least in principle. Indeed, if an exemplar is sought among the top five religions, Buddhism sometimes is singled out as a less dogmatic, less absolutist philosophy.

The third argument for violence in Religion is behavioral and comes by way of group dynamics, a field of psychology we touched on during the discussion of relative and absolute morals. Group dynamics studies the formation of groups and group behavior. In simple terms, groups form when individuals share common interests, experiences, or values. A psychological bond develops among members of the group, leading to group behaviors that sometimes differ, both in character and intensity, from how the same individuals would act if outside the group. These in-group behaviors, which nearly universally promote the interests of the group, frequently exhibit positive characteristics. Yet the negative characteristics are more pertinent to our narrative. Leadership and interpersonal dynamics often compel in-group members to feel superior to non-members (out-group members), deprecating the status and value of the latter. In-group members might even perceive out-group members as a threat. When this occurs, it becomes easier for in-group members to dehumanize and mistreat out-group members. This can occur in any type of group - in sports, politics, professions, school cliques - and can lead to racism, sexism, and other forms of discrimination directed at out-group members. It is worth noting that the German psychologist Kurt Lewin (1890-1947) is credited with putting the field of group dynamics on a scientific footing. Subsequent research validates the field.

Religion sometimes is singled out as a special case of group dynamics, owing to the intense emotional connection it inspires among its followers. This argument asserts that Religion employs its five sacreds to inspire trust and a sense of community among its members, the in-group. The five sacreds help to form the boundaries of the in-group and define interactions among its members. This bond, however, does not extend to the out-group, the non-believers. To promote and defend the in-group, its members can be persuaded to view non-believers as less vested in the goals of the in-group, less worthy of fair treatment, and less deserving of protection.[340] Negative labels (infidel, barbarian) and attributes (dirty, unintelligent) are ascribed to out-group members, further denigrating their status and value. In this manner, Religion can inspirit its followers to fanatical hatred and ruthless acts against out-group members, as evidenced by an interminable number of vile instances: Catholics butchering Protestants, Sunnis slaughtering Shiites, Christians exterminating Jews, Hindus burning Muslims, in country after country, in century after century. Religion often then accredits its most heinous acts, as well as the fealty of their perpetrators, to the supremacy and truth of their faith. This handy but repugnant stratagem offers an ancillary benefit: directing aggression at out-group members drives cohesion within the in-group.

While the assumptions and conclusions of group dynamics are well-accepted by anthropologists, psychologists, and sociologists, the same individuals are quick to acknowledge that violence stemming from in-group behavior is not confined to Religion. Secular nationalism, ethnic rivalries, and political struggles offer the same, horrific examples of in-group brutality toward out-group members. Recall the acts of Hitler, Stalin, and Mao in the twentieth century. An agnostic view might suggest that violent in-group behavior toward out-group members originates not from Religion, ethnicity, nationalism, or politics; it is a fundamental expression of our inherently tribal species to preserve a particular hierarchy and social order.

Diagram 9.3 summarizes the three arguments presented thus far. Do the rationales, either individually or collectively, call forth a conclusion? From Religion, yes. Rogue actors, not Religion itself, promote violence. It is the lack of religious faith and the absence of a divinely-inspired moral compass that precipitate violence. As case in point, religionists sometimes contend that the godlessness of the despots mentioned above - Hitler, Stalin, and Mao - led to their atrocious acts. This assertion belies a deeper truth, one that frames the stance of Reason. Let's dissect that truth.

One might rightly observe that Adolf Hitler, born into a religious household, was baptized into the Catholic Church. So, too, was Benito Mussolini. Joseph Stalin was baptized into the Russian Orthodox Church and, for several years, attended a religious seminary. Mao Zedong, some say, practiced Buddhism early in life. Yet it

Diagram 9.3
Three of the arguments suggesting that Religion promotes violence.

The Argument	The Rationale
① Scripture	Scriptural passages that advocate simony, rapine, slavery, torture, and murder induce zealots to commit heinous acts.
② Monotheism	Monotheism breeds an absolutist mentality, a mindset of "my truth is the only truth", inducing followers to mistreat those who fail to adopt the one undeniable faith.
③ Group Dynamics	Psychological bonds among in-group members breed a sense of superiority, leading to the disparagement and ill-treatment of out-group members.

would be absurd to suggest that each despot pursued his reign of terror in the name of Religion. They, like autocrats throughout history, committed their crimes from sundry motives, among them territorial gain, economic advantage, ethnic cleansing, racial superiority, the destruction of perceived enemies, or a sense of might equals right. A common thread, however, underlies all such motives, the thread of absolutism. Each oppressor translated his motive into an absolutist philosophy. One that averred indisputable truth. One that suppressed all dissent. One that tolerated no deviation from coveted aims. One that brooked no defiance to the will of the leader. And one that obligated the fanatical destruction of its opponents.

This, then, is Reason's response: Whether the ideology is secular (we are the master race) or religious (we are the chosen people), the result is identical. The proclamation of absolute truth, the consolidation of power into absolutist hands, the duty to punish all naysayers - all sow the seed for monstrous acts, a seed awaiting germination by an intransigent, totalitarian, or sociopathic leader. Reason thus contends it is the adoption of absolutist philosophies - any absolutist philosophy, including those from Religion - that all-too-frequently set the stage for potential atrocity.

IS THERE A "BIG TRUTH"?

The Greco-Roman world of the first and second centuries witnessed the emergence of an assortment of gnostic religious movements. The Gnostics[341] believed in the existence of a divine, hidden knowledge - a non-intellectual, experiential wisdom that would demystify the perennial questions of life. Why is there evil in the world? What is the soul? What is the nature of humanity? To the select few who could procure the knowledge, Gnosticism promised the earthly attainment of salvation.

Gnosticism illustrates an aspiration of myriad religious and secular philosophies: to discover the *big truth*, a holistic, all-encompassing, metaphysical explanation of our temporal existence, our spiritual essence, and our place in the universe. Is there such a truth? Religion's response is unambiguous and monepic. Yes. Every major religion asserts its existence. A truth revealed from its deity. A truth corroborated from scripture. A truth disclosed to obedient followers in exchange for their adherence to doctrinal codes of thought and behavior. Perhaps the truth is attainable to all the faithful. Perhaps it is accessible only to those who undertake an arduous temporal journey to a spiritual zenith. Or perhaps it is assigned to a preordained few. Regardless, it exists.

In superficial alignment with Religion, Reason also chases a big truth, a search for a theory of everything, a framework of laws describing the workings of the universe and all it contains. At the close of the nineteenth century, Reason believed it was close to such an understanding. It had determined the force of gravity, formulated the laws of thermodynamics, and consolidated the forces of electricity and magnetism. Several prominent physicists of the day, anecdotes suggest, presumed there was little else to discover - all that remained was to dot a few *i*'s and cross a few *t*'s. This ephemeral perspective reached an abrupt end in the early twentieth century. The discovery of quantum mechanics forced physicists to return to their blackboards and their laboratories, and compelled them to erect a new truth.

It probably is fair to say that Reason wants a big truth to exist. But Reason balances its zeal with caution, comprehending our species as systemic ideologists. That is, we possess an inclination to seek out and embrace an ideology, a tendency articulated in the previous chapter. Reason likewise recognizes that ideological adherence clashes with unbiased thinking. As a consequence, our doctrinaire appetites more likely coalesce not from facts and evidence but from psychological wants and desires, be they innate or acquired.

Reason's search for a big truth continues under various umbrellas. The Big Bang theory, string theory, quantum theory, molecular genetics, and cognitive science represent several of many paths. Whether a single truth exists remains uncertain. At this time, Reason possesses insufficient evidence to claim its existence. Reason can, however, offer two observations. If there is a big truth, it is unlikely to be the comforting, compassionate, human-centric truth that Religion craves. And if discovered, the truth will be based on empirical evidence and testable predictions, never on mystical or elusive claims.

CONCLUSION

From the preceding perspectives, note four trends. Each reflects a common theme of this narrative.

- The responses highlight stark distinctions between the *what to think* canons of Religion and Reason. The dichotomies, of course, derive from the disparate *how to think* processes of the two domains.
- The responses from Religion might vary by creed, but within each faith tend to be unequivocal and unchanging. Religion understands that resolute assurance appeals to our psychological need for certitude, which in turn influences our emotional security, our thirst for meaning, and our search for happiness. Our mental penchants, thus assuaged, reinforce devotion to the faith.
- At this stage of the narrative, the third trend is axiomatic. Religion articulates no detailed explanations or testable predictions to substantiate its responses. Its affirmations all derive from the five sacreds, with their emphasis on exclusive revelations, first-person protestations, scriptural supremacy, and emotion-as-truth aphorisms, each buttressed by appeals to authority.
- Reason insists on credible and defendable evidence. When the evidence is deficient or sketchy, Reason dispenses incomplete responses, each expressing caution, likelihoods, or skepticism. Reason's superior but demanding ability to discover objective truth thus extracts a price: It often is less able to allay psychological incertitude.

The stances of Religion and Reason suggest another distinction. Reason is more comfortable with uncertainty. At its purest, Reason relishes the unknown. The charting of the uncharted, the exploration of the unexplored, the discovery of the undiscovered - all invest Reason with a spirit to employ its *how to think* process in search of credible answers to unexplained phenomena. Its goal? To construct, brick by brick, from substratum to pinnacle, a comprehensive understanding of our world. To erect an edifice that embraces the entire spectrum of knowledge. One that encompasses the natural sciences, which entice with a lure of precise understanding - mathematics, physics, chemistry and, increasingly, biology and medicine. One that equally enfolds the social sciences, where the vagaries of human behavior challenge pinpoint prediction - sociology, psychology, economics, political science, and the humanities. It is undeniably more difficult to comprehend the workings of our brains, the organization of our societies, our ethics, our love of poetry, and our desire to create art. Yet here, too, Reason is advancing, propelled by new paradigms, new methodologies, extensive data collection, and the tenacity of passionate researchers. In time, Reason will provide well-grounded and defendable answers in these spheres as well. And when that day arrives, Reason will achieve Hume's vision of placing all of human psychology under empirical investigation.[342]

In this quest, Reason will never invent a deity, never ordain a divine text, and never proffer an incurious *Goddidit* as explanation of a mystery. It will never leverage a paucity of knowledge to justify dogmatic interpretations. It will accept "I don't know" as a valid, reasonable answer to any question, a response infinitely superior to mythical prophecy. And it will recall to us the well-worn maxim: All is mystery until it is explained.

MISCELLANY

The Estimates of the percentage of Native Americans who perished from Old World diseases between the sixteenth and eighteenth centuries vary by source.

The high mortality rate stemmed from multiple afflictions: tuberculosis, diphtheria, cholera, typhus, meningitis, smallpox, influenza, measles, hepatitis, encephalitis, and scarlet fever - all unknown in the Americas before 1492. The illnesses were brought not only by European explorers and immigrants but also by their animals - horses, cattle, pigs, and sheep.[343] All too often, the newcomers ascribed the loss of indigenous life to an act of God, a divine intervention designed to bestow land and resources on the new inhabitants by sweeping away native, non-Christian peoples. Witness John Winthrop (c.1587-1649), the first governor of the colony of Massachusetts, who declared, "For the Natives, they are near all dead of the smallpox, so as the Lord hath cleared our title to what we possess."[344]

As further affirmation of the correlation between divine absolutes and the desire for legislated intolerance, consider this statistic: From 1798 through 2014, members of Congress of the United States put forward 567 amendments (as measured by unique Joint Resolution Numbers; records from the seventeenth and eighteenth centuries are incomplete) to embed religious aspects in the Constitution. The proposals cite reliance on God in governmental matters; acknowledge God as the Supreme Authority; recognize the authority and law of Jesus Christ as Ruler of Nations; empower states to regulate morals; or endorse prayer in schools and other public buildings.[345] None was ratified.

The responses of Religion and Reason to several of humanity's fundamental inquiries.

Diagram 9.4

Perspectives

Question	Religion	Reason
Is there a supreme being?	Yes	No
What is the origin of the universe?	Divine Intervention	Natural Processes
What is the origin of life?	Divine Intervention	Natural Processes
Is life still evolving?	Varies	Yes
Are we alone?	Varies	No
Are morals absolute?	Yes	No
Does Religion promote violence?	No	Yes
Is there a "big truth"?	Yes	No

[X]

EPILOGUE

The most costly of all fallacies is to believe passionately in the palpably not true.[346]

H. L. MENCKEN (1880–1956) AMERICAN JOURNALIST

Let's remind ourselves that Religion comprises myriad belief systems, each with its own deity or deities. As such, any statement concerning Religion is bound to prompt exceptions. Yet our definition of Religion, as belief in one or more supernatural beings, applies to nearly all religions. And in most cases, if not all, the five sacreds form the basis of each faith's *how to think* process.

Similarly, our definition of Reason, as the objective search for knowledge, is, in the main, correct. As is the description of the scientific method, which is the basis for Reason's *how to think* process. Yet real-world complexities drive variations in approach, with techniques adapted to specific disciplines and their targets of examination.

Let's also again acknowledge that Religion, in its more earthly endeavors, provides numerous material and intangible benefits to societies around the world. Religion builds schools and hospitals, inspires art and architecture, battles corruption and injustice, eases hardship and poverty, maintains social order and morality, and provides a sense of protection and cohesion for communities. It is equally important to note that secular governments and institutions, in combination with enlightened cultures and humanistic values, confer similar benefits to the precariat and to all humanity. The primary debate between Religion and Reason resides not in these spheres, but in the search for knowledge and a philosophy of life, a search reflected in the *how to think* processes of the two domains. Because *how to think* drives *what to think,* Religion and Reason necessarily produce disparate sets of knowledge. It is this reality that forms the crux of the debate.

As we have seen, Religion's *how to think* process suffers from defects beyond number. The five sacreds impel Religion to suppress unwelcome inquiry, shun testable hypotheses, eschew observational evidence, disdain falsifiable predictions, evade independent verification, and scorn conclusions that contradict its beliefs. Religion must alter, cleanse, discredit, or dismiss distasteful evidence to sustain its fables. It must frame hypotheses on known evidence, then extol the same evidence as confirmation. It must practice natural theology and pseudoscience to ratify its yarns. It must rationalize emotion and intuition as adjudicators of external truth. It must exalt feel-goodism over objective reality. It must disregard the continual refutation of its prophecies, branding each disproof a test of faith. It must embrace confirmation bias, cherry-picking selected findings from science to substantiate its myths. It must rely on a distorted form of axiomatic, *a priori* logic to claim a cause is incredibly obvious, then unctuously proclaim its obviousness as proof of the cause. It must excel at circular logic, professing scripture as truth, then arguing truth from scripture. It must leverage the legions of inconsistencies within scripture to unearth verses that support its dogma, yet disregard inconvenient and contradictory passages. It must deliberately design its faith as untestable, yet proudly proclaim this same non-testability as a virtue of faith. It must assert the rationality of its own faith, while declaring the irrationality of every other faith. And when all else fails, it must invoke its most precious incantation to cast out the demons of Reason: It must appeal to authority, the core tenet of its ideology.

A final example comes from Aquinas, whose tome *Summa Thelogæ* we referenced in an earlier chapter. Aquinas penned another

Diagram 10.1
A summary of the *how to think* processes of Religion and Reason.

masterwork entitled *Summa contra Gentiles* (often translated as *The Compendium of Theology Against the Errors of Disbelievers*) to explain and defend Christian theology to followers of Judaism, Islām, and other non-Christian beliefs. *Summa contra Gentiles* comprises 464 chapters, each commenting on a specific tenet of theology. Among its chapters, nine contain the word "reason" in their titles, all of which appear in the first book of his four-volume treatise.[347] Extracts from these chapters illustrate the unsound logic of Religion, logic that Aquinas, working within the limited world understanding of his day, perhaps felt no less compelled to embrace. My comments appear in braces {}.

CHAPTER VII "The natural dictates of reason must be true: it is impossible to think of their being otherwise."[348] {Begins well, acknowledges the validity of Reason}

CHAPTER VI "There is no lightmindedness in assenting to truths of faith, although they are above reason."[349] {But quickly declares faith above Reason}

CHAPTER V "That it is an advantage for things that cannot be searched out by reason to be proposed as tenets of faith."[350] {Commits a quintessential fallacy of Religion by contending that if Reason cannot yet explain a phenomenon, the fable from Religion must prevail - another instance of the *God of the Gaps* argument}

CHAPTER VI "The divine wisdom, that knows all things most fully, has designed to reveal these her secrets to men, and in proof of them has displayed works beyond the competence of all natural powers."[351] {Asserts that revealed knowledge lies beyond the realm of Reason}

CHAPTER VI "This so wonderful conversion of the world to the Christian faith is so certain a sign of past miracles, that they need no further reiteration, since they appear evidently in their effects."[352]

{Employs circular logic to justify a claim, declaring the obviousness of alleged miracles as evidence of their existence}

CHAPTER IV "That it is an advantage for the truths of God, known by natural reason, to be proposed to men to be believed on faith."[353] {Posits faith over Reason as a reliable path to truth}

CHAPTER IV "Besides, in the midst of much demonstrated truth there is sometimes an element of error, not demonstrated but asserted on the strength of some plausible and sophistic reasoning that is taken for a demonstration."[354] {Asserts the possibility of error in Reason, but acknowledges no similar possibility of error in Religion}

CHAPTER VIII "Reason ... proves wholly insufficient to declare the substance of God Himself."[355] {Faults Reason for its inability to provide testable evidence of a religious claim}

CHAPTER X "Those truths are self-evident which are recognized at once, as soon as the terms in which they are expressed are known. Such a truth is the assertion that God exists: for by the name God we understand something greater than which nothing can be thought."[356] {Employs the ontological argument to proclaim the self-evident truth of a proposition}

CHAPTER XIII "Everything that is in motion is put and kept in motion by some other thing ... which moves other things without being itself in motion, and this we call God."[357] {Employs the cosmological argument, claiming objects in motion require a cause but demands no causal agent for a deity}

CHAPTER LVII "Reasoning means a lack of intuition: the divine knowledge therefore is not reasoned."[358] {At best, ignores that in Reason intuition is a legitimate embarkation point for inquiry but never a reliable destination; at worst, elevates intuition above Reason}

CHAPTER LXXXVI "Reason is absolutely assignable for the volitions of God regarding the universe: but relatively to us, we can assign it but vaguely, and, revelation apart, with much uncertainty."[359] {Not only dismisses the capability of human intelligence, but elevates revelation above Reason}

In stark contrast to Aquinas's opinions and unsubstantiated assertions, Reason possesses no immutable canon, no dogmatism of ignorance, no "trust me, it's true" revelations. It neither reifies ethereal objects nor fabricates vaporous spirits to explain away puzzling phenomena. Its body of knowledge is open to critique and revision, requiring no authority to authenticate its claims. Its assertions disregard the intelligence, charisma, fame, and hubris of the claimant. Reason forbids no questions, suppresses no ideas, and bans no interlocutors. Reason commands no equivalent to the *mutaween*, religious police who punish forbidden expression and behavior.[360] Reason proposes numerous hypotheses and theories, yet accepts "I don't know" as a valid answer. And unlike Religion's routine appeal to authority, the core tenet of Reason's *how to think* process is testable predictions, a tool of unequaled potency to discern fact from myth.

As Boyer states, "In every instance where the Church has tried to offer its own description of the world and there was some scientific alternative on the very same topic, the latter has proved better. Every battle has been lost and conclusively so."[361] Note the words *in every instance* in Boyer's statement, and consider the import of his reckoning. Religion's record of predicting and explaining events is abysmal. Its claims, when testable, are invariably disproved. Its eternal truths are routinely ephemeral. Its assertions of certain knowledge are a pretense. Revelation, prophecy, scripture, intuition, and piety simply afford no predictive or explanatory power. Yet Religion is adept at dismissing its failures. It conceals them in the guise of new revelations or buries them in the opaque convolutions of an imperceptible deity. Religion, to state it bluntly, is intellectually dishonest. Similar ignominious artifices in any other discipline would quickly bring the entire field into disrepute.

At the end of this chapter, I repeat the three principles asserted in the *Prologue*. The apologia presented throughout this text affirms the principles, positioning Reason as the only legitimate domain for discovering truth. Reason's legitimacy derives from its *how to think* process, the only *how to think* process that produces credible, defendable, and independently verifiable answers to both the compelling and mundane questions of life. From these traits, note three of Reason's myriad benefits.

PERSONAL Reason promises mental freedom. It allows the individual to exorcize the oppressive weight of theistic hegemony. It offers the self-reliance that emerges from assuming greater control over one's destiny, rather than shackling oneself to the temperamental acts of a notional being. It tenders new paradigms of thought and inquiry, well beyond the constraining dogma of divine authority. It affirms the transparency and grandeur of a fact-based world. It aspires to a social justice well beyond a faith's arbitrary assignment of one's lot in life. It values the secular traditions that interweave a community. And it magnifies the joy in family, friendships, and life itself, driven from the understanding that our time on earth is transitory.

PRACTICAL For millennia, humanity petitioned celestial entities for bountiful harvests, robust health, protection from the elements, and much more. But it was Reason, not Religion, that electrified our cities, invented childhood vaccines, preserved food through refrigeration, solved navigation at sea, devised buildings to withstand earthquakes, engineered drought-resistant crops, churned out antibiotics, conceived new means of transportation, crafted structures to prevent floods, created lenses to improve vision, subdued plagues, lengthened our lifespans, developed planet-wide communications, expanded crop yields, curtailed childhood mortality, constructed engines to ease physical labors, and initiated the exploration of the cosmos - to cite but a few of countless contributions. It is Reason that produces a constant stream of discoveries, driving myriad tangible benefits to humankind. Why? Because Reason works.

SOCIETAL Reason, from the Enlightenment onward, originated many of the principles now enjoyed in free societies worldwide. Here is a list of the more notable ones: constitutional government, liberal democracy, religious tolerance, the rule of law, equal rights before the law, freedom of thought and expression, separation of church from state, sovereignty residing with the people, individual liberty, gender equality, a universal conception of human rights, the search for objective truth, and the spread of knowledge. Religion, when not resisting the development of these principles, played only intermittent roles in their enactment. In today's world, in fact, more than one theocracy continues to actively resist - even condemn - the principles.

These, then, are the fruits of Reason's *how to think* process: innumerable personal, practical, and societal windfalls. All from a reliable compendium of *what to think* knowledge that establishes itself as a sound guidepost for individual and communal actions. All from an objective and enlightened framework that ennobles inquiry and advocates a rational worldview, one free from illusions, shibboleths, and superstitions. And all from a philosophy that not only propels the betterment of our societies, but cherishes a resplendent world of wonder, discovery, and hope. All this from an intelligent and pragmatic edifice entitled Reason.

Let's close this coda with a word of counsel. On your life's journey, be curious. Be a skeptic. Employ critical thinking. Question a claim with vigor proportional to the conviction of the claimant. Avoid hasty judgment. Base your actions on secular moral values. Marvel at the ineffable beauty of the cosmos and our capability to understand it. Stand in awe of the nobler virtues of our species. Treasure friends and family. Above all, enjoy life, and remember:

The universe, and our place in it,
can be explained by observation and experiment,
without resorting to mythical and supernatural explanations.

The three principles expressed in the *Prologue*, propelling this apologia of this text.

Diagram 10.2

Principles

1. There is an important distinction between *how to think* and *what to think.*
2. *How to think* is critically more important than *what to think.*
3. Only Reason possesses a sound *how to think* process and a credible *what to think* compendium of knowledge.

About The Author

Critique the idea, not its proponent.
SERGE BOURGAINE

This is the point in manuscripts where writers pen their motivations, cite their credentials, and describe their familial circumstances. Serge Bourgaine can offer none of these tidbits, for the simple reason that the moniker is a pen name. Why use a pseudonym? As expressed more than once in the narrative, no apologia should derive validity from the authority, position, or fame (or lack of these qualities) of its proponent. The facts, claims, and conclusions presented in this book thus stand on their own to be reflected upon and assessed by the reader. The same rationale explains the ample use of footnotes throughout the text. The reader is invited to authenticate any quotation, corroborate any fact, or test any inference.

I wish to thank Melvyn Bragg, host of the *In Our Time* podcast aired by the British Broadcasting Corporation, for his masterful stewardship of the eclectic series. Various episodes, their topics elucidated by credible experts in their fields, served as the starting point for much of the research behind this book.

Serge

Postscript

The *Postscript* furnishes commentary on the following topics.

A. On Chronology 297
B. On Critique 303
C. On TRUTH 307

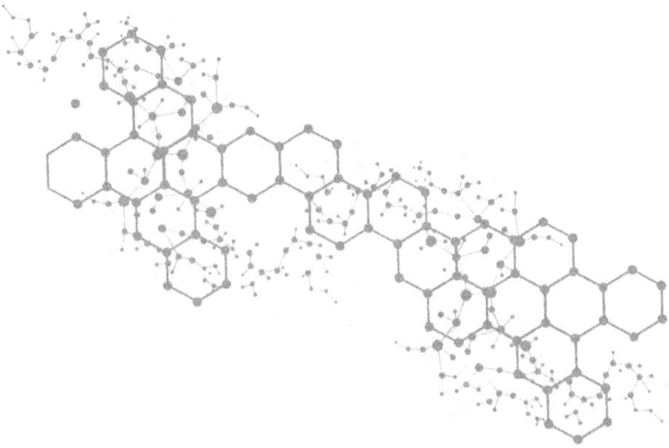

[A]

On Chronology

On communication from the Almighty:
It is revelation to the first person only, and hearsay to every other.[362]
THOMAS PAINE (1737–1809) ENGLISH POLITICAL ACTIVIST

A narrative that encompasses events from multiple civilizations over multiple millennia requires a consistent chronology. As such, this book observes both common and simplified conventions to reflect the chronological divisions of human history.

CALENDARS

Civilizations have constructed assorted calendars, many still observed to this day, to record religious and secular events. The calendars typically are based on phases of the moon (lunar calendars), movements of the sun (solar calendars), or a combination of the two (lunisolar calendars). Each calendar counts its years from a notable event. A common example is the Gregorian calendar, which counts its years from the purported birthdate of Jesus Christ (present-day scholarship suggests, however, that Jesus was born around 4 BCE). The Gregorian calendar is a solar calendar, established in 1582 under the auspices of the Catholic Church. It is the most widely used calendar in the world.

Another example is the Hijrī calendar, often referred to as the Islāmic calendar, which was established in 622 and is used to date events in various Muslim countries. Year 1 on the Hijrī calendar, which commemorates Muhammed's migration from Mecca to Medina, equates to year 622 CE on the Gregorian calendar. Because the Hijrī is a lunar calendar and the Gregorian is a solar calendar, years do not align on a one-to-one basis. The year 2020 on the Gregorian calendar, for example, covers parts of the years 1441 and 1442 on the Hijrī calendar. The Hijrī calendar also "drifts" against the Gregorian calendar by about ten days per solar year.

A third example is the Hebrew calendar, a lunisolar calendar used to determine Jewish religious observances. The Hebrew calendar counts its years from the beginning of creation. The year 2020 on the Gregorian calendar covers parts of the years 5780 and 5781 on the Hebrew calendar.

A fourth and final instance derives from Buddhism. In 1888 Thailand adopted the Gregorian calendar for legal purposes, but counts its years from the birth of Buddha, 543 years before the Christian era. The year 2020 on the Gregorian calendar, therefore, is 2563 on the Thai solar calendar. Examples of these and other extinct and extant calendars appear in Diagram 11.1.

All dates in this book are expressed on the Gregorian calendar using the BCE / CE date convention. Years in Before Common Era are marked BCE. Years in Common Era are marked only when further clarity is warranted. Aristotle was born in 384 BCE. Albert Einstein was born in 1879.

PERIODS, ERAS & AGES

To measure lengthier time spans of human activity, archaeology replaces calendars with periods, eras, and ages. Archaeological naming conventions tend to follow Greek roots: *paleo,* for example, denotes *old; lithic, stone; meso, middle;* and *neo, new.* Major divisions relevant to this book appear below. The dates of each division vary by source and geography. Not all civilizations, for example, mastered the production of bronze at the same time. For narrative simplicity, in the main text I refer to three broad eras as depicted in Diagram 11.2.

LOWER PALEOLITHIC Began 3.0 to 3.5 million years ago, marking the earliest evidence of stone tools among hominins.

Diagram 11.1
A comparison of selected calendars, both defunct and contemporary.

Calendar	Type	Year
Gregorian	Solar	2020
Egypt (Ancient)	Lunisolar	4908 /4909
French Repulic	Lunisolar	228 /229
Hebrew	Lunar	5780 / 5781
Hirjī	Solar	1441 / 1442
Holocene	Solar	12020
Indian Saka	Solar	1941 / 1942
Julian	Solar	2019 / 2020
Persian Hirjī (Iran)	Solar	1398 / 1399
Thai	Solar	2563

MIDDLE PALEOLITHIC Coincides with the rise of humans.

UPPER PALEOLITHIC Reflects advances in stone tool technology and a diversity of artifacts. Hunter-gatherers populated the Upper, Middle and Lower Paleolithic.

MESOLITHIC Reflects greater advances in stone tool technology and some early agriculture.

NEOLITHIC Marks the rise of agriculture. The Paleolithic, Mesolithic, and Neolithic comprise the Stone Age.

BRONZE AGE Signifies the widespread use of the copper and tin alloy, as well as the creation of the first early writing systems.

IRON AGE Named after the ability to smelt iron metal from iron ore, and the consequent use of the metal in weapons, tools, and structures.

POST IRON AGE Varies by geography. Sources cite various historical events to mark the end of the Iron Age, among them: the birth of the Hellenistic period in Greece, the Babylonian conquest of Jerusalem, and the rise of the Christian era.

MISCELLANY

Two principal methods exist to measure a year by the sun's movements. A solar year measures the interval between two returns of the sun to the same equinox, while a sidereal year measures the period between two returns of the sun in relation to a celestial object, typically a star. A sidereal year is approximately 20 minutes longer than a solar year. Though not technically correct, for simplicity I use the term *solar calendar* to refer to years measured by both the solar and sidereal methods.

Simplified archaeological periods. Diagram 11.2

Archaeological Periods

Archaeological Period	Time Line (Years Ago)	Simplified Periods
Middle Paleolithic	200,000 to 50,000	
Upper Paleolithic	50,000 to 20,000	Paleolithic
Mesolithic	20,000 to 10,000	
Neolithic	10,000 to 5,500	
Bronze Age	5,500 to 3,200	Neolithic
Iron Age	3,200 to 2,000	
Post Iron Age	2,000 to Today	Modern

[B]

On Critique

There are no forbidden questions in science.[363]
CARL SAGAN (1934–1996) AMERICAN PHYSICIST

Is it acceptable to critique Religion? Are some ideologies above critique? Does critique of Religion imply criticism of its followers? Is this book expressly anti-Religion? The response to each of these four questions is unequivocal: yes, no, no, no. Here's why.

It is crucial to distinguish individuals from ideologies. Individuals hold opinions and beliefs. From an admittedly Western view, this is their right, a fundamental prerogative not to be diluted, restricted, or annulled. In the United States, its Constitution guarantees this right. Each individual has a right to express their opinions and practice their beliefs. This is a core tenet of liberal democracies, which often legislate free expression and freedom of religion as protected rights.

Now consider the object of one's belief, what one believes ***in***, the ideology that one embraces. Beliefs and ideologies are not protected. They do not enjoy the same rights as the individual. Each ideology must articulate and substantiate its claims. Each must compete in the marketplace of ideas. Each wins or loses, gaining or forfeiting adherents, based on the merits of its philosophy.

Note the key distinction. Criticism of an individual differs from critique of an ideology. Christians, Muslims and atheists are individuals. Christianity, Islām, and atheism are ideologies. Criticism of individuals rarely is helpful, and seldom yields a fruitful result. Ideologies, in contrast, demand critique. They demand transparent investigation. And they demand public debate, even ridicule. This is the means by which the validity of an ideology is assessed.

Reason excels at assessing ideas and ideologies. Its *how to think* process effectively sorts the valid from the invalid, and the authentic from the illusory. It matters not whether the ideology is political, social, economic, religious, or any other. Reason demands the principles, claims, and motivations of the ideology be placed under the microscope, to be minutely examined and objectively assessed. The confirmation or refutation of an ideology derives from Reason's power of inquiry, from its demand for empirical evidence, from its reliance on testable predictions, from its insistence on independent confirmation, and from its unparalleled ability to refute spurious claims. It is an endless process, iterated generation after generation, each reconfirming valid truths and exposing beguiling misbeliefs.

To thwart the power of Reason, many societies elect to legislate against inquiry and the free expression of ideas. They attempt to enshrine into law a specific belief, proclaiming its truth and its allegedly benevolent effect in maintaining an orderly society. They tolerate no dissent from the prescribed dogma, punishing those who deviate from the imposed ideology.[364] This behavior, to practitioners of Reason, is both contemptible and despotic. Yet it also is informative. It signals weakness in the ideology. It semaphores an inability of the dogma to withstand objective scrutiny. It conveys an imperative to shield professed claims from challenge and refutation. And it communicates a necessity to sustain alleged truths by mandate of temporal authority. These acts, of course, stand in stark contrast to Reason, which demands the rigorous and continual assessment of all ideologies, including those from Reason itself. No ideology receives a free pass.

From the preceding paragraphs, it should be clear that Reason does not uniquely target Religion. Admittedly, it might appear otherwise, as revelation, scripture, and intuition-as-truth occupy no venerable locus in Reason. They receive no preferred dispensation. Yet Reason favors no ideology over another. It holds all ideologies to the same high bar: testable predictions based on objective evidence. The content of this book evinces this theme, its critiques targeted not at individuals but at ideological belief absent of evidence. As such, this text is not expressly anti-Religion. It is expressly anti-claims-without-evidence.

MISCELLANY

In the author's opinion, the right to freedom of (and freedom from) religion is eclipsed only by the rights of free speech and freedom of the press. Freedom of religion entails freedom of belief, freedom of worship, and freedom to change one's faith. These rights are enshrined in the founding documents of enlightened nation states and supranational organizations. The wording varies from document to document, but all consider freedom of religion a fundamental human right. And they are correct to do so. To cite several:

- *The Declaration of the Rights of Man and of the Citizen* (France, Article 10, 1789)
- *The Constitution of the United States of America* (Amendment 1, 1791)
- *The Australian Constitution* (Section 116, 1901)
- *The Universal Declaration of Human Rights* (United Nations, Article 18, 1948)
- *The Basic Law for the Federal Republic of Germany* (Article 3, 1949)
- *The European Convention for the Protection of Human Rights and Fundamental Freedoms* (The European Union, Article 9, 1953)

[C]

On Truth

The ideal subject of totalitarian rule is not the convinced adherent of the far-right or the far-left, but people for whom the distinction between fact and fiction, and the distinction between true and false, no longer exists.[365]

HANNAH ARENDT (1906–1975) GERMAN-AMERICAN POLITICAL THEORIST

An endless number of individuals and organizations attempt to influence your *what to think* viewpoints: religious leaders, politicians, civil servants, scientists, educators, economists, advertisers, marketers, activists, physicians, musicians, filmmakers, public relations spokespersons, legitimate news organizations, fake news organizations, law enforcement officials, businesses, bloggers, Wikipedia contributors, and social media devotees, among still more. Many offer competing and incompatible perspectives. How can one navigate an environment of conflicting information to ascertain the soundness of a claim, assess the validity of an argument, or evaluate a proposed course of action?

The practitioners of Reason proceed in three steps. First, be a skeptic; await reasonable evidence before forming an opinion. Second, proceed via inquiry, ranking observational evidence above abstract logic, ingrained beliefs, and opinions. Rely on expert commentary from reputable sources. And third, reach an objective conclusion based on a preponderance of evidence and act accordingly, always respecting your fellow citizens.

To assist inquiry, employ Reason's *how to think* repository of techniques. Consider questions from TRUTH for evaluating a claim.

Test the Claim

- Is the claim clearly articulated?
- Is the claim expressed in language free from emotion and bias?
- Is the claim supported by reasonable evidence?
- What methods were used to generate the evidence? Are the methods repeatable?
- Does the evidence include sufficient detail or omit pertinent detail?
- Does the evidence appear to be supported by sound science?
- Does the evidence assert a direct cause-and-effect relationship, a statistical likelihood, or merely a correlation?
- Does the claim offer testable and falsifiable predictions?
- What evidence would be required to refute the claim?
- Is the evidence available for examination by third parties?
- Was the claim subject to peer review?
- Does the claim cite trustworthy sources?
- Does the claim seem reasonable based on the evidence?
- Is the claim consistent with other studies? Novel findings are not necessarily incorrect, but require a higher degree of confirming evidence.
- Who benefits from the claim? Who is harmed?
- Is the claim articulated to provoke happiness and anger, or to play on emotions and belief, rather than convey objective information?

Reveal the Claimant

- Does the claimant possess credentials denoting expertise in the field?
- Does the claimant possess a record of providing verified claims?
- Does the claimant cite their organizational affiliations?
- Does the claimant remain focused on the issue, articulating

its merits and acknowledging its shortcomings?
- Does the claimant differentiate between evidence-based conclusions and personal opinions?
- Does the claimant appear to advocate a particular outcome, rather than following wherever the evidence might lead?
- Does the claimant dismiss contradictory evidence?
- Does the claimant appeal to authority to substantiate the claim?
- Does the claimant rely on hyperbole, false analogy, misinformation, or disinformation to advance a claim?
- Does the claimant respond calmly and rationally to challenges, or become irate and feign indignation?
- Does the claimant attempt to advance a claim based on their reputation, prestige, position, or vanity?
- Does the claimant attempt to discredit, humiliate, or shout down opponents?
- Does the claimant resort to personal attacks and calumnies to advance a claim?
- Does the claimant profess to be the only one who can discern the truth?
- Does the claimant appear likely to accept evidence that contradicts his or her existing beliefs?
- Is it possible to contact the claimant?

UNDERSTAND THE ORGANIZATION

- Who funded the research?
- Who published the claim?
- How old is the publishing organization? Is the organization reputable, with a long history of responsible journalism?
- Do organizations that comment on the claim offer balanced perspectives on a variety of topics?
- Do organizations that comment on the claim operate an internal process to vet the reasonableness of claims, or do they merely parrot commentary from other organizations?

- Do organizations that comment on the claim appear to have been established explicitly to champion a specific worldview?
- Do organizations that comment on the claim appear to ridicule opponents of their worldview?

Typify the Society

- Does the society tolerate open inquiry?
- Does the society legislate against undesirable opinions on circumscribed topics?
- Does the society empower official groups to police adherence to a prescribed set of viewpoints?
- Does the society tolerate non-official groups who censor objectionable viewpoints and maltreat their proponents?
- Does the society punish proponents of opposing viewpoints, thereby creating a culture of self-censorship?

Heighten Your Self-Awareness

- Do you understand your biases?
- Do you seek diverse viewpoints from a variety of sources?
- Do you live in an information bubble, educating yourself from a small number of sources that reinforce previously-held convictions?
- Do you attempt to dismiss uncomfortable evidence?
- Do you prefer to base decisions heavily on emotion and gut instincts?
- Are you open to revising your opinions in light of compelling new evidence?
- Are you willing, on topics of significance to you, to undertake further study to understand the methods and reliability of research?

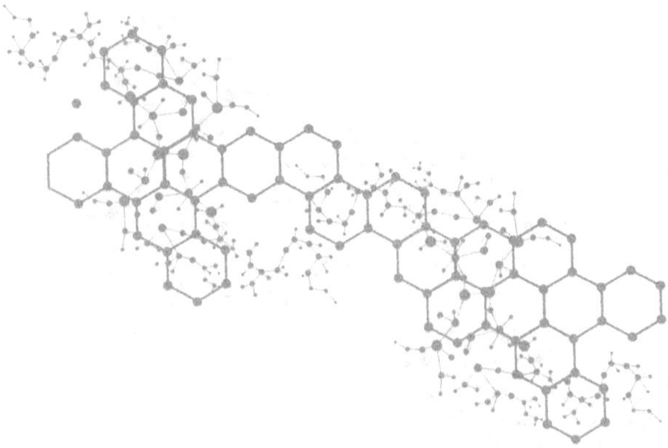

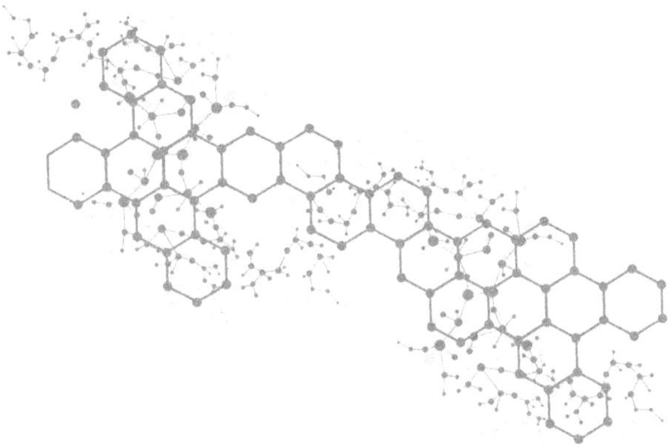

Endnotes

Entries that omit page numbers refer, in nearly all instances, to ebooks or similar digital content.

1 Wilde 1905, p. 188.
2 Batson 1993, p. 77.
3 King James Bible Online, John 12:17.
4 Kohn 2007, not paginated.
5 King James Bible Online, Genesis 4:15.
6 Historical Department Journal 1849, not paginated.
7 Taylor 1884, p. 304.
8 Bush 1984, not paginated.
9 The Doctrine and Covenants, Official Declaration 2.
10 Mead 1970, p. 72. This widely quoted truism appears to be a rephrasing. The likely source text: *We must create new models for adults who can teach their children not what to learn, but how to think, and not what they should be committed to, but the value of commitment*.
11 The Royal Society.
12 Hugo 1877, p. 236. Source text: *Il y a maintenant en France dans chaque village un flambeau allumé, le maître d'école, et une bouche qui souffle dessus, le curé*.
13 Breuilly 2005, p. 16.
14 Durkheim 1964, p. 47.
15 Bowker 1997, p. xv.
16 Geertz 1973, pp. 90-91.
17 Marx 2014, p. 3. Source text: *Die Religion ist Seufzer der bedrängten Kreatur, das Gemüt einer herzlosen Welt, wie sie der Geist geistloser Zustände ist. Sie ist das Opium des Volks*.
18 Bragg, History as Science, 11 March 1999.
19 Castor 2006, not paginated.
20 Diamond 2005, pp. 83+.
21 Lord 2016, not paginated.
22 Kohn 2004, not paginated.

[23] Fix 2005b, not paginated.
[24] Tuchman 2014a, not paginated.
[25] Esposito 1999, p. 84.
[26] Bragg, Sunni and Shia Islam, 25 June 2009.
[27] Breuilly 2005, p. 69.
[28] Bragg, Epic of Gilgamesh, 3 November 2016.
[29] Pseudo-Apollodorus 2016, not paginated.
[30] Bragg, The Trinity, 13 March 2014.
[31] King James Bible Online, Psalms 5:3.
[32] King James Bible Online, 1 Thessalonians 5:17.
[33] Breuilly 2005, p. 69.
[34] Breuilly 2005, pp. 59,79,101,117.
[35] Bragg, Prophecy, 13 June 2013.
[36] Bragg, The Delphic Oracle, 30 September 2010.
[37] Fix 2005a, not paginated.
[38] King James Bible Online, 2 Timothy 3:16.
[39] King James Bible Online, Proverbs 2:10.
[40] The Hadīth, Sunan Ibn Majah, 1:68.
[41] Herman 1991, p. 120.
[42] Bhaktivedanta 1989, p. 322.
[43] Walters 2002, p. 30.
[44] Tuchman 2014a, not paginated.
[45] Hobbes 2003, p. 85.
[46] Asimov 1981, not paginated.
[47] Aristotle 2003a, Bekker 25a1+, pp. 39+
[48] Hume 1960, p. 124.
[49] Pierce 1901, not paginated.
[50] Duclaux 1896, p. 142. *Source text: On voit bien ici de quelle série de jugements sans cesse revisés se fait le progrès incontestable de la science.*
[51] Singh 2005, p. 496.
[52] Smith 2003, p. 390.
[53] Sagan 1990, not paginated.
[54] Atkinson 1974, pp. 123-131.
[55] Bragg, Pliny's Natural History, 8 July 2010.
[56] Bragg, Humanism, 8 February 2001.
[57] Bragg, The Translation Movement, 2 October 2008.
[58] Bragg, The Abbasid Caliphs, 2 February 2006.
[59] The Hadīth, Sunan Ibn Majah, 1:229.
[60] Al-Khalili 2011, not paginated.
[61] Bragg, al-Biruni, 10 June 2010.
[62] Bragg, Al-Kindi, 28 June 2011.

[63] Bragg, Avicenna, 8 November 2007.
[64] Al-Khalili 2011, not paginated.
[65] Al-Khalili 2011, not paginated.
[66] Bragg, Muslim Spain, 22 November 2002.
[67] Bragg, Baconian Science, 2 April 2009.
[68] Bacon 2003, p. 108.
[69] Aristotle 2003b, Bekker 249b30-250a5, p. 333.
[70] Farmelo 2011, not paginated.
[71] Whewell 1847, pp. 60-74.
[72] Popper 2002, p. 17.
[73] Batson 1993, p. 379.
[74] Cronin 1991, p. 32.
[75] Feynman 2012, Special Preface.
[76] Elsberry 2007, p. 252.
[77] Bragg, Empiricism, 10 June 2004.
[78] Tyson 2019, p. 25.
[79] Feynman 1994, p. 150.
[80] Gribbin 2003, pp. 511-512.
[81] Stevin 1955, pp. 508-511.
[82] Stevin 1955, p. 510. Source text: *Aristoteles voornomde eueredenheyt is onrecht*.
[83] Newton 2003, p. 8.
[84] Newton 2003, proposition 75, theorem 35, p. 134.
[85] Feynman 1994, p. 14.
[86] Einstein 2003, p. 228.
[87] Singh 2005, p. 121.
[88] Singh 2005, pp. 125-126.
[89] Singh 2005, p. 132.
[90] Bragg, Gravitational Waves, 17 May 2007.
[91] Kuhn 1970, p. 112.
[92] Animal Factory 2016, pp. 105-107.
[93] Al-Khalili 2011, not paginated.
[94] Mencken 1956, aphorism 412, p. 277.
[95] Matthew 2001, p. 133.
[96] Whewell 1847, aphorism xvi, p. 560.
[97] Sagan 2000, timestamp 00:01:25.
[98] Galilei 2003, pp. 157-160, 208.
[99] Sagan 1980, pp. 14-15.
[100] Singh 2005, p. 14.
[101] Ptolemy 2003, pp. 9-12.
[102] Norlind 1953, pp. 273-276.
[103] Luther 1914, p. 126. Source text: *Ratio ... ist die hochste hur, die der*

Teuffel hat.

104 Galilei 1880, not paginated. Galileo's discovery of moons in orbit around Jupiter is mentioned in the dedication to Cosmo II, the Grand Duke of Tuscany, with further detail provided in later pages of *The Sidereal Messenger*.

105 Bragg, The Spanish Inquisition, 22 June 2006.

106 Singh 2005, pp. 60-75.

107 Stephen 1890, pp. 253-254.

108 The Word on the Street 2004, not paginated.

109 Tuchman 2014a, not paginated.

110 Al-Khalili 2011, not paginated.

111 Al-Khalili 2011, not paginated.

112 Barnhart 2015, not paginated.

113 Barnhart 2015, not paginated.

114 Domonoske 2016, not paginated.

115 Azhar 2017, not paginated.

116 Saudi Blogger Badawi 2015, not paginated.

117 Singapore Jails Teen 2017, not paginated.

118 Kopan 2016, not paginated.

119 King James Bible Online, Daniel 8:14.

120 Nichol 1945, p. 12.

121 Bragg, St Thomas Aquinas, 17 September 2009.

122 Bragg, Logic, 21 October 2010.

123 Aristotle 2003a, Bekker 25a1+, pp. 39+. Also Aristotle 2003c, Bekker 21a1, p. 32.

124 Aristotle 2003d, Bekker 296b25, p. 388.

125 Aristotle 2003d, Bekker 289a20, p. 380.

126 Aristotle 2003d, Bekker 289b31, p. 381.

127 Aristotle 2003d, Bekker 269a30, p. 360.

128 Aquinas 2003, pp. 346-7.

129 Aquinas 2003, pp. 354–5.

130 Aquinas 2003, p. 346.

131 Aquinas 2003, p. 177.

132 Aquinas 2003, pp. 12-13.

133 King James Bible Online, Joshua 10:12-13.

134 King James Bible Online, Exodus 14:21.

135 King James Bible Online, 2 Kings 2:8.

136 King James Bible Online, Mark 5:11-13.

137 King James Bible Online, Jonah 2:1-10.

138 King James Bible Online, 2 Kings 6:5-6.

139 King James Bible Online, Exodus 8:1-14.

140 Pope Gregory I 1911, pp. 232-233.

[141] The Noble Qur'an, 54:17.
[142] Sagan 1996, not paginated.
[143] Sagan 1996, not paginated.
[144] Russell 1997, pp. 547-8.
[145] Bury 1913, p. 20.
[146] Drake 1978, pp. 162,165.
[147] Bragg, The Scientist, 24 October 2002.
[148] Aristotle 2003d, Bekker 286b10, p. 378.
[149] Aristotle 2003d, Bekker 297b25, p. 389.
[150] Aristotle 2003e, Bekker 1073a30, p. 603.
[151] Aquinas 2003, p. 350.
[152] Aquinas 2003, p. 356.
[153] Dobzhansky 1945, p. 75.
[154] Bragg, The Age of the Universe, 3 March 2011.
[155] Dawkins 2009, not paginated.
[156] Dawkins 2009, not paginated.
[157] Dawkins 2009, not paginated.
[158] Dawkins 2009, not paginated.
[159] Dalrymple 1991, pp. 125+.
[160] Bragg, The Earth's Origins, 5 July 2001.
[161] Russell 1913, p. 76.
[162] Bragg, The Earth's Origins, 5 July 2001.
[163] Dalrymple 1991, p. 53.
[164] Boltwood 1907, pp. 78-88.
[165] Holmes 1913, pp. 155-157.
[166] Bragg, Maimonides, 17 February 2011.
[167] Sagan 1980, p. 258.
[168] Duncan 1998, pp. 1-6.
[169] Duncan 1998, p. 72.
[170] Duncan 1998, pp. 133-135.
[171] Duncan 1998, p. 194.
[172] Bragg, Early Geology, 12 April 2012.
[173] Burnett 1719, p. 10.
[174] Burnett 1719, pp. 10-17.
[175] King James Bible Online, Genesis 7:12.
[176] Burnett 1719, p. 157.
[177] Burnett 1719, p. 10.
[178] King James Bible Online, Psalms 9:17.
[179] The Noble Qur'an, 4:140.
[180] Singh 2013, not paginated.
[181] Wesley 1999, sermon 129, not paginated.
[182] The Noble Qur'an, 10:37.

[183] King James Bible Online, Matthew 12:30.
[184] King James Bible Online, Mark 10:9.
[185] Tagliaferro 2010, not paginated.
[186] Damer 1980, pp. iii-v.
[187] Pascal 1904, pp. 149-150. Source text: *Mais il y a ici une infinité de vie infiniment heureuse à gagner, un hasard de gain contre un nombre fini de hasards de perte, et ce que vous jouez est fini.*
[188] Mill 2003, p. 300.
[189] Dawkins 2016, not paginated.
[190] Singh 2005, p. 4.
[191] Bragg, Hindu Ideas of Creation, 5 December 2003.
[192] Mayr 1982, p. 336.
[193] Darwin 1878, p. 380.
[194] Darwin 2003, pp. 18-23.
[195] Darwin 2003, pp. 63-64.
[196] Bragg, Alfred Russel Wallace, 21 March 2013.
[197] Dennett 2017, p. 82.
[198] Gould 2007, not paginated.
[199] Bragg, Fossils, 22 March 2001.
[200] Bragg, Fossils, 22 March 2001.
[201] Bragg, Ediacara Biota, 9 July 2009.
[202] Bragg, The Cambrian Period, 17 February 2005.
[203] Gould 2007, not paginated.
[204] Dawkins 2009, not paginated.
[205] Bragg, Plate Tectonics, 24 January 2008.
[206] Dawkins 2009, not paginated.
[207] Abell 1983, pp. 35-36.
[208] Dalrymple 2007, pp. 170.
[209] Foxx 2016, not paginated.
[210] Mendez 2013, p. 454.
[211] Shoemaker 1983, p. 489.
[212] Williams 2000, p. 56.
[213] Suyu 2010, p. 219.
[214] Lunardini 1995, p. 20.
[215] Liu 2000, p. 186.
[216] Ford 1999, p. 88.
[217] Dawkins 2009, not paginated.
[218] Dawkins 2009, not paginated.
[219] Dawkins 2009, not paginated.
[220] Ruse 1982, p. 308.
[221] Evolution: Birds Introduced to Hawaii Evolved in Mere Decades 2019, p. 16.

[222] Dawkins 2009, not paginated.

[223] Dawkins 2009, not paginated.

[224] Bragg, Genetics, 13 December 2001.

[225] Dawkins 2009, not paginated.

[226] Dawkins 2009, not paginated.

[227] Bragg, Human Evolution, 16 February 2016.

[228] Bragg, The Neanderthals, 17 June 2010.

[229] Cronin 1991, pp. 36-39.

[230] Dawkins 2009, not paginated.

[231] Cronin 1991, pp. 45-46.

[232] Cronin 1991, p. 12.

[233] Augustine 1836, p. 218. Source text: *Quapropter bono Christiano, sive mathematici, sive quilibet impie divinantium, maxime dicentes vera, cavendi sunt, ne consortio dæmoanimam deceptam, pacto quodam societatis irretiant*.

[234] Duncan 1998, p. 157.

[235] Pope Pius IX 1864, error 8.

[236] Pope Pius IX 1870, session 4.

[237] Pope Benedict XVI 2009, article 56.

[238] Catechism of the Catholic Church 2003, paragraph 159.

[239] Singh 2013, not paginated.

[240] Sastri 1898, p. 49.

[241] The Hadīth, Sunan Ibn Majah, 1:269.

[242] King James Bible Online, Romans 14:23.

[243] Molden 2005, pp. 295-312.

[244] Edwards 1972, p. 203.

[245] Dawkins 2009, not paginated.

[246] Gould 1995, pp. 5-8.

[247] George 2009, not paginated.

[248] Krauss 2015, not paginated.

[249] Boyer 2001, p. 297.

[250] Gould 2007, not paginated.

[251] Bragg, Ockham's Razor, 31 May 2007.

[252] Jung 1923, p. 569.

[253] Reich 2018, not paginated.

[254] Festinger 2014, not paginated.

[255] Aristotle 2003f, Bekker 1102b25, p. 348.

[256] Maslow 1943, pp. 370-396.

[257] Reiss 2015, pp. 1-10.

[258] Teehan 2010, not paginated.

[259] King James Bible Online, Genesis 1:27.

[260] King James Bible Online, Genesis 1:26.

261 King James Bible Online, Psalms 37:4.
262 King James Bible Online, Proverbs 3:6.
263 King James Bible Online, John 3:36.
264 Keats 1909, not paginated.
265 Vatican Condemns 2016, not paginated.
266 Catechism of the Catholic Church 2003.
267 Pinker 1997, pp. 556-557.
268 Tuchman 2014b, not paginated.
269 Allen 2001, p. 340.
270 Calvin 1845b, p. 534.
271 Calvin 1845b, p. 541.
272 Calvin 1845a, p. 46.
273 Calvin 1845a, p. 188.
274 Beit-Hallahmi 1997, pp. 11, 96, 112-113, 162, 242.
275 Boyer 2001, p. 47.
276 Bush 2017, not paginated.
277 King James Bible Online, Psalms 8:2. Note that the 4/14 Window Movement cites scripture not from the King James Bible but from the Holman Christian Standard Bible.
278 Pope John Paul II 2002, not paginated.
279 Jakarta's Governor: Demolition in Progress 2017, p. 35.
280 Jakarta Governor Ahok Found Guilty of Blasphemy 2017, not paginated."
281 Plato 2003, Stephanus 414, p. 340.
282 Plato 2003, Stephanus 415, p. 341.
283 Plato 2003, Stephanus 429, p. 347.
284 Gibbon 2003, p. 12. Also attributed to Seneca the Younger.
285 Teehan 2010, not paginated.
286 Batson 1993, pp. 128-129.
287 Dimitropoulos 2017, pp. 26-28.
288 Bragg, Lamarck and Natural Selection, 26 December 2003.
289 Gould 1995, pp. 41-57.
290 Al-Khalili 2011, not paginated.
291 McLean v Arkansas Board of Education 1982.
292 Edwards v Aguillard 1987.
293 Kitzmiller v Dover Area School District 2005.
294 Tobin 2012b, timestamp 00:05:40.
295 Hooke 1665, observation 44, p. 194.
296 Festinger 1962, pp. 1-3, 47.
297 Festinger 1962, pp. 259-264.
298 Festinger 2014, not paginated.
299 King James Bible Online, Matthew 7:15.
300 The Noble Qur'an, 10:35.

[301] Keysar 2014, pp. 553-582.
[302] Hitchens 2007, p. 150.
[303] King James Bible Online, Genesis 35:11.
[304] Bragg, Ontological Argument, 27 September 2012.
[305] Hume 1907, pp. 116-122.
[306] Kant 2003, pp. 179-182.
[307] Bragg, Science and Religion, 25 January 2001.
[308] Edwards 1972, p. 204.
[309] King James Bible Online, Genesis 1:1.
[310] The Noble Qur'an, 7:54.
[311] Singh 2013, not paginated.
[312] Tobin 2012a, timestamp 00:16:30.
[313] Dorminey 2019, not paginated.
[314] Rees 2000, pp. 52-57.
[315] Bragg, Grand Unified Theory, 24 February 2000.
[316] Bragg, The Origin of Infectious Diseases, 9 June 2011.
[317] Weintraub 2014, pp. 81, 117, 133, 139, 141, 144, 157, 167, 174, 184, 191, 197, 205.
[318] The Doctrine and Covenants, 76:24.
[319] Sagan 1980, p. 299.
[320] Scoles 2011, pp. 1-4.
[321] Diamond 2005, pp. 268-269.
[322] Teehan 2010, not paginated.
[323] Veenhoven 2007, not paginated.
[324] The Freedom of Thought Report 2015, p. 1+.
[325] Rousseau 1912, p. 81. Source text: *Le chef-d'œuvre d'une bonne éducation est de faire un homme raisonnable: et l'on prétend élever un enfant par la raison! C'est commencer par la fin, c'est vouloire faire l'instrument de l'ouvrage. Si les enfants entendaient raison, ils n'auraient pas besoin d'être élevés.*
[326] Teehan 2010, not paginated.
[327] King James Bible Online, Numbers 31:17-18.
[328] King James Bible Online, Deuteronomy 22:13-21.
[329] King James Bible Online, 1 Samuel 6:19.
[330] King James Bible Online, Luke 19:27.
[331] Pope John Paul II 1997, not paginated.
[332] The Noble Qur'an, 9:5.
[333] The Noble Qur'an, 22:19.
[334] The Hadīth, Sahih Muslim, 1:33.
[335] The Hadīth, Sahih Muslim, 54:105.
[336] The Hadīth, Sahih al-Bukhari, 88:5.
[337] The Noble Qur'an, 4:34.

338 The Hadīth, Bulugh al-Maram, 10:1209.
339 Bragg, Baltic Crusades, 24 November 2016.
340 Teehan 2010, not paginated.
341 Bragg, Gnosticism, 2 May 2013.
342 Hume 1960, pp. xvii-xxiii.
343 Mann 2011, not paginated.
344 Emerson 1976, p. 116.
345 Amending America 2016, not paginated.
346 Mencken 1949, p. 616.
347 Aquinas 1905. The specified count corresponds to the Rickaby translation.
348 Aquinas 1905, p. 28.
349 Aquinas 1905, p. 27.
350 Aquinas 1905, p. 25.
351 Aquinas 1905, p. 27.
352 Aquinas 1905, p. 27.
353 Aquinas 1905, p. 23.
354 Aquinas 1905, p. 23.
355 Aquinas 1905, p. 30.
356 Aquinas 1905, p. 32.
357 Aquinas 1905, p. 36.
358 Aquinas 1905, p. 95.
359 Aquinas 1905, p. 137.
360 Saudi Arabia's Religious Police 2016, p. 41.
361 Boyer 2001, p. 320.
362 Paine 1896, p. 23.
363 Sagan 1996, not paginated.
364 Rousseau 2003, p. 399.
365 Arendt 1962, p. 474.

Works Cited

The bibliography is organized by author surname and publication date.

Abell, George O. 1983. The Ages of the Earth and the Universe. *Scientists Confront Creationism*. Edited by Laure R. Godfrey. New York: W. W. Norton.

Al-Khalili, Jim. 2011. *The House of Wisdom: How Arabic Science Saved Ancient Knowledge and Gave Us the Renaissance*. New York: Penguin.

Allen, Charles. 2001. *Soldier Sahibs: The Men Who Made the North-West Frontier*. London: Little Brown.

Amending America: Proposed Amendments to the United States Constitution, 1784 to 2014. 2016. The U. S. National Archives and Records Administration. [archives.gov]

Animal Factory: The Evolution of a Scientific Mainstay. 2016. *The Economist*. 24 December 2016.

Aquinas, Thomas. 1905. *Of God and His Creatures: An Annotated Translation of The Summa Contra Gentiles of St Thomas Aquinas*. First published circa 1265. Translated by Joseph Rickaby. London: Burns & Oates.

Aquinas, Thomas. 2003. *The Summa Theologica, Volume 1*. First published 1485. Great Books of the Western World, Volume 17. Edited by Mortimer J. Adler, translated by Father Shapcote of the Fathers of the English Dominican Province, seventh printing. Chicago: Encyclopædia Britannica.

Arendt, Hannah. 1962. *The Origins of Totalitarianism*. Seventh edition. New York: Meridian.

Aristotle. 2003a. *Prior Analytics*. First published circa 350 BCE. Great Books of the Western World, Volume 7. Edited by Mortimer J. Adler, translated by G. R. G. Mure, seventh printing. Chicago: Encyclopædia Britannica.

Aristotle. 2003b. *Physics*. First published circa 350 BCE. Great Books of the Western World, Volume 7. Edited by Mortimer J. Adler, translated by R. P. Hardie and R. K. Gaye, seventh printing. Chicago: Encyclopædia Britannica.

Aristotle. 2003c. *On Interpretation*. First published circa 350 BCE. Great Books of the Western World, Volume 7. Edited by Mortimer J. Adler, translated by E. M. Edghill, seventh printing. Chicago: Encyclopædia Britannica.

Aristotle. 2003d. *On the Heavens*. First published circa 350 BCE. Great Books of the Western World, Volume 7. Edited by Mortimer J. Adler, translated by J. L. Stocks, seventh printing. Chicago: Encyclopædia Britannica.

Aristotle. 2003e. *Metaphysics*. First published circa 350 BCE. Great Books of the Western World, Volume 7. Edited by Mortimer J. Adler, translated by W. D. Ross, seventh printing. Chicago: Encyclopædia Britannica.

Aristotle. 2003f. *Nicomachean Ethics*. First published circa 350 BCE. Great Books of the Western World, Volume 8. Edited by Mortimer J. Adler, translated by W. D. Ross, seventh printing. Chicago: Encyclopædia Britannica.

Asimov, Isaac. 1981. The Threat of Creationism. *The New York Times*. 14 June 1981.

Atkinson, R. J. C. 1974. Neolithic Science and Technology. *Philosophical Transactions of the Royal Society*. Volume 276, issue 1257, 2 May 1974.

Augustine. 1836. *De Genesi Ad Litteram*. First published 415 CE. Collectio Selecta SS Ecclesiæ Patrum. Edited by D. A. B. Caillau. Parisiis: Apud Parent-Desbarres.

Azhar, Mobeen. 2017. Pakistan's Secret Atheists. *BBC Magazine*. 12 July 2017.

Bacon, Francis. 2003. *Novum Organum*. First published 1620. Great Books of the Western World, Volume 28. Edited by Mortimer J. Adler, seventh printing. Chicago: Encyclopædia Britannica.

Barnhart, Edwin. 2015. Maya Hieroglyphs: Breaking the Code. *Maya to Aztec, Ancient Mesoamerica Revealed*. The Great Courses, course 3100, lecture 13. Chantilly: The Teaching Company. [video]

Batson, C. Daniel. 1993. *Religion and the Individual: A Social-Psychological Perspective*. New York: Oxford University Press.

Beit-Hallahami, Benjamin, and Michael Argyle. 1997. *The Psychology of Religious Behaviour, Belief and Experience*. London: Routledge.

Bhaktivedanta, A. C. 1989. *Bhagavad-Gītā As It Is*. Second edition. Alachua: Bhaktivedata Book Trust.

Boltwood, Bertram B. 1907. On the Ultimate Disintegration Products of the Radioactive Elements. *American Journal of Science*. Series 4, volume 23, number 134, February 1907, pp. 78-88.

Bowker, John, editor. 1997. *The Oxford Dictionary of World Religions*. Oxford: Oxford University Press.

Boyer, Pascal. 2001. *Religion Explained: The Evolutionary Origins of Religious Thought*. New York: Basic.

Bragg, Melvyn. *In Our Time*. British Broadcasting Corporation (BBC). [audio podcast]

Breuilly, Elizabeth, et al. 2005. *Religions of the World: The Illustrated Guide to Origins, Beliefs, Traditions & Festivals*. Oxford: Transedition and Fernleigh Books.

Burnett, Thomas. 1719. *The Sacred Theory of the Earth*. First published 1681. Fourth edition. London.

Bury, J. B. 1913. *A History of Freedom of Thought*. New York: Henry Holt.

Bush, Lester E., and Arnold L. Mauss, editors. 1984. *Neither White Nor Black: Mormon Scholars Confront the Race Issue in a Universal Church*. Midvale, Signature Books.

Bush, Luis. 2017. *4/14 Movement*. [414movement.com]

Calvin, John. 1845a. *The Institutes of the Christian Religion, Volume 1*. First published 1536. Translated by Henry Beveridge. Edinburgh: Calvin Translation Society.

Calvin, John. 1845b. *The Institutes of the Christian Religion, Volume 2*. First published 1536. Translated by Henry Beveridge. Edinburgh: Calvin Translation Society.

Castor, Alexis Q. 2006. Farming and Early Settlements. *Between the Rivers: The History of Ancient Mesopotamia*. The Great Courses, course 3180, lecture 5. Chantilly: The Teaching Company. [video]

Catechism of the Catholic Church. 2003. First published 1992. The Holy See. [vatican.va]

Cronin, Helena. 1991. *The Ant and the Peacock: Altruism and Sexual Selection from Darwin to Today*. Cambridge: Cambridge University Press.

Dalrymple, G. Brent. 1991. *The Age of the Earth*. Stanford: Stanford University Press.

Dalrymple, G. Brent. 2007. The Ages of the Earth, Solar System, Galaxy and Universe. *Scientists Confront Creationism: Intelligent Design and Beyond*. Edited by Andrew J. Petto and R. Godfrey Laurie. New York: W. W. Norton.

Damer, Edward T. 1980. *Attacking Faulty Reasoning*. Belmont: Wadsworth.

Darwin, Charles. 1878. *Journal of Researches into the Natural History and Geology of the Counties Visited During the Voyage of H.M.S. Beagle Round the World, Under the Command of Capt. Fitz Roy, R. N.* New York, D. Appleton.

Darwin, Charles. 2003. *The Origin of Species by Means of Natural Selection*. First published 1859. Great Books of the Western World, Volume 49. Edited by Mortimer J. Adler, seventh printing. Chicago: Encyclopædia Britannica.

Dawkins, Richard. 2009. *The Greatest Show on Earth: The Evidence for Evolution*. New York: Free Press.

Dawkins, Richard. 2016. *The Selfish Gene: Fortieth Anniversary Edition*. First published 1976. Fortieth Anniversary Edition. Oxford: Oxford University Press.

Dennett, Daniel. 2017. *From Bacteria to Bach and Back: The Evolution of Minds*. New York: W. W. Norton.

Diamond, Jared. 2005. *Guns, Germs and Steel: The Fates of Human Societies*. New York: W. W. Norton.

Dimitropoulos, Stav. 2017. Trying to Lose My Religion. *Discover*. September 2017.

Dobzhansky, Theodosius. 1945. Evolution, Creation and Science. *The American Naturalist*. Volume 79, number 780, January-February 1945.

The Doctrine and Covenants. 2013. The Church of Jesus Christ of Latter Day Saints. [churchofjesuschrist.org]

Domonoske, Camila. 2016. *Atheist Law Student Hacked to Death in Bangladesh*. National Public Radio (NPR). 7 April 2016. [audio transcript]

Dorminey, Bruce. 2019. Without the Moon, Would There Be Life on Earth? *Scientific American*. 21 April 2009.

Drake, Stillman. 1978. *Galileo at Work: His Scientific Biography*. New York: Dover.

Duclaux, Émile. 1896. *Pasteur: Histoire d'un Esprit*. Sceau: Imprimerie Charaire.

Duncan, David Ewing. 1998. *Calendar: Humanity's Epic Struggle to Determine a True and Accurate Year*. New York, Avon.

Durkheim, Émile. 1964. *The Elementary Forms of the Religious Life*. First published 1912. Translated by Joseph Ward Swain. London: George Allen & Unwin.

Edwards, Paul. 1972. *The Encyclopedia of Philosophy, Volume 1*. First published 1967. New York: Macmillan.

Edwards v Aguillard. 1987. United States Supreme Court. Justice William Brennan, citation 482 U.S. 578, 19 June 1987.

Einstein, Albert. 2003. *The General Theory of Relativity*. First published 1915. Great Books of the Western World, Volume 56. Edited by Mortimer J. Adler, translated by Robert W. Lawson, seventh printing. Chicago: Encyclopædia Britannica.

Elsberry, Wesley. 2007. Logic and Math Turn to Smoke and Mirrors: William Dembski's Design Inference. *Scientists Confront Creationism: Intelligent Design and Beyond*. Edited by Andrew J. Petto and Laurie R. Godfrey. New York: W. W. Norton.

Emerson, Everett, editor. 1976. John Winthrop to Sir Nathanial Rich, 22 May 1634. *Letters from New England: The Massachusetts Bay Colony, 1619-1638*. Amherst: The University of Massachusetts Press.

Esposito, John L., editor. 1999. *The Oxford History of Islam*. New York: Oxford University Press.

Evolution: Birds Introduced to Hawaii Evolved in Mere Decades. 2019. *New Scientist*. Issue 3230, 18 May 2019.

Farmelo, Farmelo. 2011. *The Strangest Man: The Hidden Life of Paul Dirac, Mystic of the Atom*. New York: Basic Books.

Festinger, Leon. 1962. *A Theory of Cognitive Dissonance*. First published 1957. Stanford: Stanford University Press.

Festinger, Leon, Henry W. Riecken, and Stanley Schachter. 2014. *When*

Prophecy Fails: A Social and Psychological Study of a Modern Group that Predicted the Destruction of the World. Radford: Wilder Publications.

Feynman, Richard. 1994. *The Character of Physical Law*. First published 1965. New York: Modern Library.

Feynman, Richard. 2012. *Six Easy Pieces: Essentials of Physics Explained by Its Most Brilliant Teacher*. First published 1962. New York: Basic Books.

Fix, Andrew C. 2005a. The Hundred Years' War and the Church in Crisis. *The Renaissance, the Reformation and the Rise of Nations*. The Great Courses, course 3940, lecture 2. Chantilly: The Teaching Company. [video]

Fix, Andrew C. 2005b. Martin Luther. *The Renaissance, the Reformation and the Rise of Nations*. The Great Courses, course 3940, lecture 16. Chantilly: The Teaching Company. [video]

Ford, Derek C., and Carol A. Hill. 1999. Dating of Speleothems in Kartchner Caverns, Arizona. *Journal of Cave and Karst Studies*. Volume 61, number 2, August 1999, pp. 84-88.

Foxx, Chris. 2016. *Australia Plans New Coordinates to Fix Sat-Nav Gap*. British Broadcasting Corporation (BBC). 29 July 2016.

The Freedom of Thought Report. 2015. International Humanist and Ethical Union (IHEU).

Galilei, Galileo. 1880. *The Sidereal Messenger*. First published 1610. Translated by Edward Stafford Carlos. London: Rivingtons.

Galilei, Galileo. 2003. *Dialogues Concerning the Two New Sciences*. First published 1638. Great Books of the Western World, Volume 26. Edited by Mortimer J. Adler, translated by Henry Crew and Alfonso de Salvio, seventh printing. Chicago: Encyclopædia Britannica.

Geertz, Clifford. 1973. *The Interpretation of Cultures: Selected Essays*. New York: Basic Books.

George, Alison. 2009. Eye-Burrowing Worms, National Treasures, and Creationism. *New Scientist*. 28 January 2009.

Gibbon, Edward. 2003. *The Decline and Fall of the Roman Empire, Volume 1*. First published 1776. Great Books of the Western World, Volume 37. Edited by Mortimer J. Adler, seventh printing. Chicago: Encyclopædia Britannica.

Gould, Stephen Jay. 1995. *Adam's Navel and Other Essays*. London: Penguin.

Gould, Stephen Jay. 2007. *Wonderful Life: The Burgess Shale and the Nature of History*. First published 1989. New York: W. W. Norton.

Gribbin, John. 2003. *Science: A History*. London: Penguin Books.

The Hadīth. First published circa 632 CE. [sunnah.com]

Herman, A. L. 1991. *A Brief Introduction to Hinduism*. Boulder: Westview Press.

Historical Department Journal History of the Church, 1830-2008, Volume 26. 1849. The Church of Jesus Christ of Latter-Day Saints. Church History Library. Call number CR 100 137, 13 February 1849.

Hitchens, Christopher. 2007. *God is Not Great: How Religion Poisons*

Everything. New York: Twelve / Hachette.

Hobbes, Thomas. 2003. *Leviathan*. First published 1651. Great Books of the Western World, Volume 21. Edited by Mortimer J. Adler, seventh printing. Chicago: Encyclopædia Britannica.

Holmes, Arthur. 1913. *The Age of the Earth*. London: Harper & Brothers.

Hooke, Robert. 1665. *Micrographia: Some Physiological Descriptions of Minute Bodies made by Magnifying Glasses with Observations and Inquiries Thereupon*. London, Jo. Marten and Ja. Allestry.

Hugo, Victor. 1877. *Histoire d'un Crime: Déposition d'un Témoin*. Fourth edition. Paris: Ancienne Maison Michel Lévy Frères.

Hume, David. 1907. *Dialogues Concerning Natural Religion*. First published 1779. Edinburgh: William Blackwood.

Hume, David. 1960. *A Treatise of Human* Nature. First published 1740. Oxford: Clarendon Press.

Jakarta Governor Ahok Found Guilty of Blasphemy. 2017. British Broadcasting Corporation (BBC). 9 May 2017.

Jakarta's Governor: Demolition in Progress. 2017. *The Economist*. 28 January 2017.

Jung, Carl. 1923. *Psychological Types*. First published 1921. England: T. and A. Constable.

Kant, Immanuel. 2003. *The Critique of Pure Reason*. First published 1781. Great Books of the Western World, Volume 39. Translated by J. M. D. Meiklejohn, edited by Mortimer J. Adler, seventh printing. Chicago: Encyclopædia Britannica.

Keats, John. 1909. Keats: *Poems Published in 1820*. Oxford, Clarendon Press.

Keysar, Ariela, and Juhem Navarro-Rivera. 2014. A World of Atheism: Global Demographics. *The Oxford Handbook of Atheism*. Edited by Stephen Bullivant and Michale Ruse. Oxford: Oxford University Press.

King James Bible Online. First published 1611. [kingjamesbibleonline.org]

Kitzmiller v Dover Area School District. 2005. United States District Court, M.D. Pennsylvania. Justice John Jones III, citation 400 F. Supp. 2d 707, 20 December 2005.

Kohn, Rachel. 2004. God's Wife. *The Ark*. 2 May 2004. Australia Broadcasting Corporation (ABC). [audio transcript]

Kohn, Rachel. 2007. Mormonism's Lost Tribes. *The Ark*. 25 March 2007. Australia Broadcasting Corporation (ABC). [audio transcript]

Kopan, Tal, and Eugene Scott. 2016. *North Carolina Governor Signs Controversial Transgender Bill*. Cable News Network (CNN). 24 March 2016.

Krauss, Lawrence M. 2015. Ben Carson's Scientific Ignorance. *The New Yorker*. 28 September 2015.

Kuhn, Thomas S. 1970. *The Structure of Scientific Revolutions*. First published

1962. International Encyclopedia of Unified Science, Volume 2, Number 2. Second edition. Chicago: University of Chicago Press.

Liu, Tanzhou, and Wallace S. Broecker. 2000. How fast does rock varnish grow? *Geology*. Volume 11, number 2, February 2000, pp. 183-186.

Lord, John. 2016. *A Short Introduction to Ancient Religions*. Endymion Press.

Lunardini, Virgil J. 1995. *Permafrost Formation Time*. Cold Regions Research and Engineering Laboratory (CRREL). US Army Corps of Engineers. Report 95-08, April 1995.

Luther, Martin. 1914. *Text der Predigten des Jahres 1546*. First published 1549. D. Martin Luthers Werke: Kritischer Gesamtausgabe, Volume 51. Sermon number 3, 17 January 1546. Weimar: Herman Böhlaus Nachfolger.

Mann, Charles C. 2011. *1491: New Revelations of the Americas Before Columbus*. Second edition. New York, Vintage.

Marx, Karl. 2014. *Zur Kritik der Hegelschen Rechtsphilosophi*. First published 1844. Edited by Michael Holzinger. Berlin: CreateSpace Independent Publishing Platform.

Maslow, Abraham. 1943. A Theory of Human Motivation. *Psychological Review*. Volume 50, issue 4, July 1943.

Matthew, Edmond A. 2001. *Earth: Inside and Out*. New York: New Press.

Mayr, Ernst. 1982. *The Growth of Biological Thought: Diversity, Inheritance and Evolution*. Cambridge: Harvard University Press.

McLean v Arkansas Board of Education. 1982. United States District Court for the Eastern District of Arkansas. Justice William Overton, citation 529 F. Supp. 1255, 5 January 1982.

Mead, Margaret. 1970. *Culture and Commitment*. New York: Natural History Press / Doubleday.

Mencken, H. L. 1949. *A Mencken Chrestomathy*. New York: Alfred A. Knopf.

Mencken, H. L. 1956. *Minority Report: H. L. Mencken's Notebooks*. Baltimore: John Hopkins University Press.

Mendez, Fernando L. 2013. An African American Paternal Lineage Adds an Extremely Ancient Root to the Human Y Chromosome Phylogenetic Tree. *The American Journal of Human Genetics*. Volume 92, 7 March 2013, pp. 454-459.

Mill, John Stuart. 2003. *On Liberty*. First published 1859. Great Books of the Western World, Volume 40. Edited by Mortimer J. Adler, seventh printing. Chicago: Encyclopædia Britannica.

Molden, Daniel C., and E. Tory Higgins. 2005. Motivated Thinking. *The Cambridge Handbook of Thinking and Reasoning*. Edited by K. Holyoak and B Morrison. New York: Cambridge University Press.

Newton, Isaac. 2003. *Mathematical Principles of Natural Philosophy*. First published 1687. Great Books of the Western World, Volume 32. Edited by Mortimer J. Adler, translated by Andrew Motte and Florian Cajori, seventh

printing. Chicago: Encyclopædia Britannica.
Nichol, Francis D. 1945. *The Midnight Cry: A Defense of the Character and Conduct of William Miller and Millerites, Who Mistakenly Believed that the Second Coming of Christ Would Take Place in the Year 1844*. Washington, D.C.: Review and Herald Publishing.
The Noble Qur'an. First published circa 632 CE. [quran.com]
Norlind, Wilhelm. 1953. Copernicus and Luther: A Critical Study. *Isis*. Volume 44, number 3, September 1953.
Paine, Thomas. 1896. *The Age of Reason*. First published 1794-1807. The Writings of Thomas Paine, Volume 4. Edited by Moncure Daniel Conway. New York: G. P. Putnam's Sons.
Pascal, Blaise. 1904. *Pensées, Volume 2*. First published 1670. Paris: Hachette.
Pierce, Charles Sanders. 1901. *On the Logic of Drawing History from Ancient Documents, Especially from Testimonies*. Collected Papers of Charles Sanders Pierce. Manuscript MS [R] 690, CP 7.218. Commens Digital Companion to C. S. Pierce.
Pinker, Steven. 1997. *How the Mind Works*. London: W. W. Norton.
Plato. 2003. *The Republic*. First published circa 380 BCE. Great Books of the Western World, Volume 6. Edited by Mortimer J. Adler, translated by Benjamin Jowett, seventh printing. Chicago: Encyclopædia Britannica.
Pope Benedict XVI. 2009. *Caritas in Veritate*. Papal Encyclicals Online. 29 June 2009.
Pope Gregory I. 1911. *The Dialogues of Saint Gregory the Great*. First published circa 590 CE. Edited by Edmund G. Gardner, book 4, chapter 39. London: Phillip Lee Warner.
Pope John Paul II. 1997. *Address of His Holiness Pope John Paul II to a Symposium on the Roots of Anti-Judaism*. Delivered 31 October 1997. The Holy See.
Pope John Paul II. 2002. *Canonization of Juan Diego Cuauhtlatoatzin*. Papal Encyclicals Online. 31 July 2002.
Pope Pius IX. 1864. *Syllabus of Errors*. Papal Encyclicals Online. 8 December 1864.
Pope Pius IX. 1870. *Decrees of the First Vatican Council*. Papal Encyclicals Online. Session 4, 18 July 1870.
Popper, Karl. 2002. *The Logic of Scientific Discovery*. First published 1959. London: Routledge Classics.
Pseudo-Apollodorus. 2016. *The Library of Apollodorus*. First published circa 100 CE. Translated by James George Frazer. Hastings: Delphi Classics.
Ptolemy. 2003. *The Amalgest*. First published circa 150 CE. Great Books of the Western World, Volume 15. Edited by Mortimer J. Adler, seventh printing. Chicago: Encyclopædia Britannica.
Rees, Martin. 2000. *Just Six Numbers: The Deep Forces that Shape the*

Universe. New York: Basic Books.

Reich, David. 2018. *Who We Are and How We Got Here: Ancient DNA and the New Science of the Human Past*. New York, Pantheon.

Reiss, Steven. 2015. *The 16 Strivings for God: The New Psychology of Religious Experiences*. Macon: Mercer University Press.

Rousseau, Jean-Jacques. 1912. *Émile, Volume 1*. First published 1762. Paris: Jean Gillequin.

Rousseau, Jean-Jacques. 2003. *The Social Contract*. First published 1762. Great Books of the Western World, Volume 35. Edited by Mortimer J. Adler, translated by G. D. H. Cole, seventh printing. Chicago: Encyclopædia Britannica.

The Royal Society. [royalsociety.org]

Ruse, Michael. 1982. *Darwinism Defended: A Guide to the Evolution Controversies*. Reading: Addison-Wesley.

Russell, Alexander. 1913. *Lord Kelvin: His Life and Work*. London, T. C. & E. C. Jack.

Russell, Bertrand. 1997. *Last Philosophical Testament 1943-68*. The Collected Papers of Bertrand Russell, Volume 11. Edited by John G. Slater. London, Routledge.

Sagan, Carl. 1980. *Cosmos*. New York: Random House.

Sagan, Carl. 1990. Why We Need to Understand Science. *Skeptical Inquirer*. Volume 14, issue 3.

Sagan, Carl. 1996. *The Demon-Haunted World: Science as a Candle in the Dark*. New York: Ballantine.

Sagan, Carl. 2000. Encyclopedia Galactica. *Cosmos: A Personal Voyage*. Episode 12. Aired by Public Broadcasting Service (PBS). [video]

Sastri, A. Mahadeva. 1898. *Amritabindu and Kaivalya Upanishads with Commentaries*. Madras, Thomson.

Saudi Arabia's Religious Police: Advice for the Vice Squad. 2016. *The Economist*. 22 October 2016.

Saudi Blogger Badawi Flogged for Insult to Islam. 2015. British Broadcasting Corporation (BBC). 9 January 2015.

Scoles, Sarah, and Sue Ann Heatherly. 2011. 50 Years of Giving Direction to the Scientific Search for Life Beyond Earth. *The Universe in the Classroom*. Number 77, Winter 2011. Astronomical Society of the Pacific.

Shoemaker, Eugene M. 1983. Asteroid and Comet Bombardment of the Earth. *Annual Review of Earth and Planetary Sciences*. Volume 11, May 1983, pp. 461-494.

Singapore Jails Teen Blogger Amos Yee for Anti-Religion Posts. 2016. British Broadcasting Corporation (BBC). 29 September 2016.

Singh, Simon. 2005. *Big Bang*. First published 2004. Hammersmith: Harper Perennial.

Singh, Gobind. 2013. *Shri Guru Granth Sahib*. Khalsa Consensus Translation. Global Grey.

Smith, Adam. 2003. *An Inquiry Into the Nature and Causes of the Wealth of Nations*. First published 1776. Great Books of the Western World, Volume 36. Edited by Mortimer J. Adler, seventh printing. Chicago: Encyclopædia Britannica.

Stephen, Leslie, and Sidney Lee, editors. 1890. *Dictionary of National Biography, Volume 24*. Entry on Matthew Hamont written by Alexander Gordon. New York: Macmillan .

Stevin, Simon. 1955. *General Introduction: Mechanics*. First published 1586. The Principle Works of Simon Stevin, Volume 1. Edited by E. J. Djiksterhuis. Amsterdam: C. V. Swets & Zeitlinger.

Suyu, S. H. 2010. Dissecting the Gravitational Lens B1608+656. *The Astrophysical Journal*. Volume 711, 1 March 2010, pp. 201-221.

Tagliaferro, Edward P. 2010. Mother Teresa on Divorce. *Times of Malta*. 19 October 2010.

Taylor, John. 1882. *Journal of Discourses, Volume 22*. Brigham Young University Digital Collections. Liverpool: Albert Carrington.

Teehan, John. 2010. *In the Name of God: The Evolutionary Origins of Religious Ethics and Violence*. Chichester: John Wiley & Sons.

Tobin, Harold J. 2012a. The Early Ocean and the Origins of Life. *Oceanography: Exploring Earth's Final Wilderness*. The Great Courses, course 1730, lecture 9. Chantilly: The Teaching Company. [video]

Tobin, Harold J. 2012b. Tides in Theory and Practice. *Oceanography: Exploring Earth's Final Wilderness*. The Great Courses, course 1730, lecture 16. Chantilly: The Teaching Company. [video]

Tuchman, Barbra Barbara W. 2014a. *A Distant Mirror: The Calamitous Fourteenth Century*. First published 1978. New York: Random House.

Tuchman, Barbara. 2014b. *The Proud Tower: A Portrait of the World Before the War, 1890-1914*. First published 1966. New York: Random House.

Tyson, Neil deGrasse. 2019. *Letters From an Astrophysicist*. New York: W. W. Norton.

Vatican Condemns Radio Station Over Anti-Gay Comments on Quake. 2016. British Broadcasting Corporation (BBC). 4 November 2016.

Veenhoven, Ruut. 2007. *Measures of Gross National Happiness*. Presented at OECD World Forum, Istanbul, 30 June 2007.

Walters, J. Donald. 2002. *Intuition for Starters*. Nevada City, Crystal Clarity.

Weintraub, David A. 2014. *Religions and Extraterrestial Life: How Will We Deal With It?* Springer Praxis Books.

Wesley, John. 1999. *The Cause and Cure of Earthquakes*. First published 1750. The Sermons of John Wesley. Wesley Center Online.

Whewell, William. 1847. *The Philosophy of the Inductive Sciences, Founded*

Upon Their History, Volume 2. First published 1840. London: John W. Parker.

Wilde, Oscar. 1905. The Critic as Artist, Part II. *Intentions*. New York: Brentano's.

Williams, George E. 2000. Geological Constraints on the Precambrian History of Earth's Rotation and the Moon's Orbit. *Reviews of Geophysics*. Volume 38, 1 February 2000, pp. 37-59.

The Word on the Street. 2004. National Library of Scotland. First published 1826. L. C. Folio 73(093a).

De omnibus dubitandum est

ORBIS
UNUM
PRESS

An imprint of La Jolie Vie, LLC

Made in United States
Troutdale, OR
11/03/2026

56845106R00204